Clinical Ethics on Film

M. Sara Rosenthal

Clinical Ethics on Film

A Guide for Medical Educators

 Springer

M. Sara Rosenthal
Program for Bioethics, Departments of
   Internal Medicine, Pediatrics and
   Behavioral Science
University of Kentucky
Lexington
USA

ISBN 978-3-030-08001-3   ISBN 978-3-319-90374-3   (eBook)
https://doi.org/10.1007/978-3-319-90374-3

This Springer imprint is published by the registered company Springer International Publishing AG
part of Springer Nature
The registered company address is: Gewerbestrasse 11, 6330 Cham, Switzerland

To my maternal grandparents, Jacob Lander, M.D. (1910–1989), who would have especially loved the film, Awakenings; and Clara Lander, Ph.D. (1916–1978), a John Donne scholar, who would have especially loved the film, Wit.

Preface

The Marriage of Clinical Ethics and Film

Clinical Ethics is a subspecialty of bioethics that deals with ethical dilemmas "at the bedside" that specifically involves the healthcare provider–patient relationship or sometimes the healthcare provider's moral distress over a particular dilemma. Such dilemmas comprise end of life care; weighing therapeutic benefits against risks and side effects; clinical trials and experimental therapies; harms to patients or patient populations caused unintentionally or by deliberate acts; medical error; healthcare access and healthcare discrimination; and medical professionalism and integrity.

There are a myriad of films that demonstrate the nuances and dilemmas in clinical ethics; some notable "clinical ethics" films in this collection, such as *Wit* (2001) and *Whose Life is it, Anyway?* (1981) originated from stage plays. Some, such as *The Elephant Man* or *Awakenings* (1990) originated from nonfiction memoir. Still others were conceived as original screenplays.

This book arose from years of work I had done in the area of clinical ethics pedagogy using film to teach medical students about clinical ethics principles and dilemmas. With limited human resources in my small bioethics program, as well as limited classroom time available to medical students and residents for ethics teaching, I began a long project of creating several online teaching modules that used over 100 film clips in documentary-styled modules, each with an approximate running time of an hour. Ultimately, this work grew into a popular elective bioethics course called "Bioethics on Film" for our healthcare trainees, including medical students, nursing students, and allied health students. Lester Friedman, Professor of English and Humanities, was one of the first to document the pedagogic utility of showing films to healthcare students (Friedman 1995):

> The films function as extended case histories that encourage students to see various sides of an ethical dilemma…[Film] provides an accessible point of reference that personalizes the often arcane philosophical commentaries that characterize healthcare ethics debates….The

ethics teacher stimulates students to integrate abstract principles with concrete situations, a methodology which encourages students to combine theory and practice into an organic whole.

The requirement to teach clinical ethics in academic medical centers did not evolve from the institutional level; medical school deans did not suddenly have an epiphany and decide to allocate funds for bioethics teaching. The requirement to teach this content evolved as requirements from the national councils that grant accreditation to academic medical centers. The Liaison Committee for Medical Education (LCME), which is responsible for medical school accreditation, began to require bioethics teaching in 2005. In the area of graduate medical education, since 2007, the Accreditation Council of Graduate Medical Education (ACGME) has included a set of core competencies in its Common Program Requirements for medical residents, and the requirement to teach ethics has become more rigorous. Among these competencies is professionalism, which includes the call for "an adherence to ethical principles." These required competencies mandate ethics education in all areas of medicine.

Bioethicists come from a variety of core disciplines including philosophy, law, medicine, social science, social work, religious studies, and many more. Unlike Friedman, or other film scholars, my core graduate discipline is medical sociology, which informs my theoretical frameworks and my work as a clinical ethicist; it also informs the content in the films I've selected for this collection and my discussion of the film's social and historical location. My undergraduate disciplines—literature and film—inform the sections of text devoted to the film's making and director's intentions with telling a story.

Using film in the teaching of bioethics is certainly not new, as Friedman demonstrates. Many of my colleagues have been using film in their bioethics teaching for over 20 years, and there are several interesting bioethics film courses peppering different academic centers. What is new is the creation of a "clinical ethics" film collection that is distinct from a general "bioethics film" collection. This book comprises the core clinical ethics films surrounding autonomy and beneficence.

What defines the film as a "clinical ethics" film?

The criteria for my film selection in this book are as follows: The film must involve as a central theme, either a patient's experiences with illness; a healthcare provider's experiences; or the healthcare provider–patient relationship. Thus, films with only occasional scenes or minor subplots of the above are excluded. Or, the film takes place in a healthcare setting: a hospital or academic medical center; a medical research setting; and a treatment center, such as addiction treatment facility. Or, finally, the film's central theme is about a clinical ethics issue, and all the scenes in the film contribute to this theme.

By using the above criteria, I eliminate films that are about wider bioethics or social justice and policy issues, such as *Gattaca* (1997), *Soylent Green* (1973), or *District 9* (2009).

Another important element in my selection is the quality of the film. *Is it a good film* regardless of the above criteria? Several films in this collection won critical acclaim and awards. I've divided this book into three clinical ethics genres. Part I, "End of Life On Film", comprises films that deal with death and dying. The films selected in this section comprise *Wit* (2001), *Whose Life is It, Anyway?* (1981), *My Life* (1993), and *All That Jazz* (1979).

Wit is a well-established end of life film that is shown all over the country. The film chronicles an English literature professor's experiences with being enrolled in a clinical trial for treatment of Stage IV ovarian cancer. Professor Vivian Bearing, a world-renowned scholar on the works of John Donne and metaphysical wit, experiences what she was taught by her mentor: that Donne intended a comma in his line: "And death shall be no more" comma "death, thou shalt die." She is unprepared for the medical model and medicalized death. If I were restricted to only one end of life film in this collection, *Wit* would be it, hence the chapter's title. Originally a stage play (1995), *Wit* was made into an HBO film that was never a major theatrical release, but has won critical acclaim as well as acclaim by healthcare providers who work in palliative care and bioethics. I also use *Wit* to demonstrate a variety of nursing ethics themes, which I discuss in Chap. 1.

Whose Life is It, Anyway? showcases an extraordinary cast, including the late John Cassavettes. Based on the stage play of the same name, this film serves as social commentary on the 1976 Quinlan decision, and what some trace to the early development of clinical ethics as a subspecialty of Bioethics.

My Life is a sleeper that was not particularly successful at the box office. Based on an original screenplay, this film deals with "family systems" and many of the psychosocial problems at the end of life that are caused by an absence of closure between terminally ill patients and their family members. Michael Keaton gives an understated performance of a dying man whose battle is less with his illness than with finding closure with his family members, including his unborn child.

All That Jazz (1979) is an unusual choice in a clinical ethics film collection. This musical is rarely discussed as an End of Life film, yet the entire film is exactly this. It is about a patient in denial about his reckless habits, and his dying. It happens to be an autobiography of Bob Fosse's death from heart disease. The film takes us into the patient's creative imagination and existential suffering at the end of life. This film also boasts nine Academy Award nominations, and is ranked 14 out of the American Film Institute's best 100 musicals.

Part II, entitled "Films About Competency And Decision-Making Capacity", comprises *Diving Bell and the Butterfly* (2007); *One Flew Over the Cuckoo's Nest* (1975), which also won the 1975 Academy Award for Best Picture; and *Still Alice* (2014). These are films that deal with ethical dilemmas specifically involving a patient's ability to make a medical decision and/or adequately communicate.

Diving Bell and the Butterfly is based on the autobiography of the former editor of the French-based *Elle Magazine*. The film is about a successful endeavor by French healthcare providers to find a way to communicate with a stroke patient who has lost almost all of his capacity to communicate. It also highlights stark differences in French health care and American health care.

One Flew Over the Cuckoo's Nest (1975) looks at malingering, capacity, and psychiatric harms. It is based on the novel by Ken Kesey, who worked in mental hospitals circa the late 1960s.

Still Alice deals with gradual loss of capacity in a patient diagnosed with early-onset Alzheimer's disease, for which Julianne Moore won her academy award.

Part III, entitled "Films About Beneficence", looks at films that focus on benefits versus harms in a clinical setting. All of the films selected here are based on documented cases in the medical literature: *The Elephant Man* (1980), *Lorenzo's Oil* (1992), and *Awakenings* (1990).

The Elephant Man is based on the true story of Joseph Merrick, who had a disfiguring, rare disease, resulting in his being exhibited in the infamous freak shows of the Victorian Age. He is discovered by Dr. Frederick Treaves, who gives him comfort and permanent hospice in a London Hospital. Treaves asks at one point if he is a "good man or a bad man" when faced with weighing whether he has truly maximized benefits for this patient, or merely exploited him as a freak again within the medical model.

Lorenzo's Oil is based on the true pediatrics case of Lorenzo Michael Odone who had adrenoleukodystrophy. His parents would not accept a terminal diagnosis and developed a still-controversial treatment for their son. Odone died at the age of 30, however, and is still the longest-surviving patient with adrenoleukodystrophy. The film highlights ethical issues with parental authority and decision-making.

Awakenings based on the book by Oliver Sacks is the true story of the early L-DOPA trials in which an astounding short-term benefit was seen with the drug L-DOPA on a subset of neurological patients, but harsh side effects emerged when the drug was used for long periods of time.

Readers who find value in this collection may wish to explore the companion work to this book, Healthcare Ethics on Film (2019), also comprising three sections: Part I, "Films About Medical Harms"; Part II, "Films About Distributive Justice and Healthcare Access"; and Part III, "Films about Professionalism and Humanism".

Ultimately, this book is a crossover of clinical ethics and film. It is important to note, however, this is about *clinical ethics* and not film theory or film studies. This book should be viewed as interdisciplinary. It offers rich content for both healthcare and film pedagogy. You will find the expected discussions surrounding the back-story on each film in this collection, and the various historical accidents that led to the film's making. Perhaps, more unique about this book is the analytical frameworks used to discuss the selected films: medical sociology, history of medicine, and clinical ethics. This is how I teach these films to my students, and how the films can be best understood as *clinical ethics* films. These films not only serve as commentaries on clinical ethics issues but are also timepieces that speak to a particular point in American medical and cultural history. Thus, each film in this collection is discussed within its social location and medical history context in addition to the specific clinical ethics themes expressed.

As a result, the healthcare themes of many of the films in this collection spoke differently to its original theater audience upon the film's release than to the audience watching today. What can a present lens offer to explain poor box office reception of some films in this collection? Was the film ahead of its time? Was the audience not receptive to the healthcare themes in the film? For example, a 1979 musical about a chain-smoking, pill-popping patient's experiences with heart disease is a vastly different film in a timeframe where addiction research is in its infancy, the causes of heart disease are not well understood, and smoking is a cultural norm. Similarly, a 1981 drama about a patient requesting withdrawal of life support is a shocking theme in this time frame, when the medical community was still reverberating from re: Quinlan, medical paternalism is the norm, and the concept of patient autonomy has not penetrated into mainstream medicine. At the same time, this collection offers many academy award-winning gems. I anticipate that for the clinical ethics or healthcare reader, this book contains much more content about the film itself, rather than just using the film as a vehicle to debate bioethics issues (see Shapsay (2009), for example). Out of the ten films discussed in this volume through a dedicated chapter, several are based on works of nonfiction, and can thus function as dramatic presentations of case studies. As for the fictional films, such as *Wit*, Friedman (1995) notes that for healthcare students:

> [W]hile fictional films play an important role in my ethics classes, they function mainly to motivate students to examine their feelings about a particular issue. The artistry of such works – the narrative, visual, aural and dramatic elements that coalesce to make them such effective pedagogical tools – rarely enters our discussions, since the films remain a pathway to moral dilemmas rather then focus of critical analysis: the medium is certainly not the message.

However, regardless of whether films are based in fiction or nonfiction, how a director tells a story visually is part of the story, and healthcare students may still benefit from such discussions.

For the film scholar, this book may appear to have less content about the technical aspects of the film, but will medically locate the film both historically and sociologically, which I expect will broaden a film scholar's technical analysis of the film when engaged in such discussion with students.

One frequent question I get with respect to using films in teaching is that of process: costs and the best ways for students to screen films. My process has evolved due to time considerations and student populations. I have screened films in class, but that requires at least two and a half hours of class time for a post-screening discussion. In more recent years, I have assigned the films as readings or provide "screening days" as most students prefer to screen the film on their own or in small groups, which enables me to devote the entire formal class time to discussion of the film. In some cases, the films are intense, and students prefer to screen these alone because they may trigger very strong emotions. I also provide "screening notes" to accompany the students' screenings, which helps guide their screening experience.

To avoid licensing and copyright issues in an ever-changing digital environment, I recommend that instructors provide links to *legitimate* retail sites that offer *rental streaming* for the films (such as Amazon, Netflix, Hulu, or HBOGO). At the same time, I recommend that instructors also purchase the DVD of the film and keep them on reserve at their institution's library for students to screen in case there are problems with the internet—particularly as issues such as net neutrality remain in question. In my case, although my institution will cover the cost of these DVD purchases, I prefer to have my own collection. Some of my colleagues, who have accounts with streaming sites, will simply login during class to screen certain films. Much depends on the timing and purpose of your screening. In the United States, *The TEACH Act* allows you to screen any film for educational purposes as part of Fair Use for educators. Several of the films will be available on YouTube at any given point in time, but it's best to avoid providing YouTube or other suspect websites as "official" links, although students are, of course, free to seek them out themselves. Students with financial constraints will still be able to use your library reserve copies for screening. Are all the films in this collection available for streaming rental? The answer is usually, but licensing agreements can change over time. As of this writing, most of the films in this collection have digital rental streaming options, but should some become unavailable, purchasing the hard DVD for students to screen is always an option. It is more likely in the future that DVD copies may become obsolete, but in those cases, libraries can purchase institutional streaming rights to some films that students can use.

Ultimately, this book should serve as a unique collection of clinical ethics films for the myriad of disciplines attracted to it. Whether you are planning a course on healthcare films as a film studies scholar, planning a film course as a healthcare scholar, or like to buy film books as a film buff, my hope is to provide you with a robust collection of clinical ethics films to choose from—many of which, you may never have thought of as a "clinical ethics film".

Lexington, USA   M. Sara Rosenthal

References

Friedman, Lester. 1995. See me, hear me: Using film in health-care classes. *Journal of Medical Humanities*. 16:223–228.

Shapshay, Sandra, ed. 2009. *Bioethics at the Movies*. JHU Press.

Acknowledgements

I wish to thank Kimberly S. Browning, MPH. for her assistance with research on many of the films in this book. Jeffrey P. Spike, Ph.D., Director, Campus-wide Ethics Program and Professor, McGovern Center for Humanities and Ethics provided helpful comments on early drafts of Chaps. 1 and 8. Jeffrey Tuttle, M.D., Department of Psychiatry, University of Kentucky provided helpful comments on Chap. 6. Finally, Kenneth B. Ain, M.D., Professor, Department of Internal Medicine, University of Kentucky, and also my spouse and best friend, watched endless hours of films with me, and provided helpful insights and resources surrounding this book's content.

Contents

About the Author

M. Sara Rosenthal, Ph.D. is Professor and Founding Director of the Program for Bioethics, Departments of Internal Medicine, Pediatrics and Behavioral Science, University of Kentucky, Lexington, Kentucky, USA.

Healthcare providers usually begin to identify their first clinical ethics dilemmas in patients who are dying, or at the end of life. In many of these situations, healthcare providers may have moral distress as well. Healthcare providers in the roles of mentoring and teaching may suffer from lingering moral residue themselves from such cases, which may affect the teaching culture and healthcare trainee expectations. Indeed, many healthcare trainees still do not get adequate training in end of life dialogues and truthful prognostication, which may not be available without skilled mentors. Finally, reducing unit moral distress that affects learners also requires formal educational forums for difficult end of life cases as well as an institutional mechanism for effective clinical ethics consultation and moral distress debriefings. Screening one of the films in this section as a unit exercise, followed by a panel discussion, may help serve that purpose if there are no other educational venues.

Consider each film in this section an end of life case from the *patient's* point of view, in which the Principles of Autonomy and Respect for Persons dominate the discussion. In each film, the patient's experiences, preferences, and medical circumstances challenge whether his/her end of life preferences are reasonable, are being honored, and whether the goals of care should be adjusted.

In *Wit*, which depicts a "bad death" experience, we wonder whether the patient would have agreed to aggressive treatment had she truly appreciated the risks involved with her treatment. This film raises hard questions about informed consent, advance directives, moral distress, and nursing ethics. Next, students will travel back in time to the paternalistic hospital cultures circa Quinlan and Cruzan to champion a quadriplegic's request for withdrawal of life support in *Whose Life Is It, Anyway?* Understanding the ethical dilemmas patients and their surrogates waged over prolonging death is "ground zero" in the history of clinical ethics. In *My Life*, a white guy dying of renal cell carcinoma has a "good death" where we see the impact of good hospice care and how family systems play a role in end of life. Finally, in *All That Jazz*, we examine a true story of a non-compliant "VIP" patient dying from heart disease in the late 1970s due to his bad behaviors, self-centered "Me Generation" worldview, and addictions. It challenges us to question whether he should even be a surgical candidate on beneficence grounds, but it's a patient everyone has had or will have in health care.

Core Bioethics Principles Involved

Ultimately, the core bioethics principle dominating in this section is the *Principle of Autonomy*, where a patient's preferences, consistent with his/her values and beliefs, ought to guide goals of care at the end of life. The Principle of Respect for Persons is a dual obligation to respect autonomous patient preferences, *but to also protect patients who do not have autonomy* by ensuring there is an authentic surrogate decision-maker, for example, or that previously stated preferences or advance directives are honored.

The Principle of Beneficence obligates healthcare providers to maximize clinical benefits and minimize clinical harms by typically ensuring there is a greater balance of benefits than harms, but this is not always easy to sort out. In this section, we must ask whether autonomous patient preferences and values take priority over beneficence, and when, in some cases, beneficence necessarily limits autonomy.

The Principle of Non-maleficence is the explicit obligation not to knowingly offer a treatment that has no benefit, or to harm patients; this is often a conjoined principle with Beneficence as there may not be a bright line where violating Non-Maleficence is an issue. However, in *Whose Life is it, Anyway?*, the idea of withdrawing life support was considered a violation of this principle by the healthcare provider.

Finally, the Principle of Justice in clinical ethics looks at healthcare access and resources, and when we look at the films in this section, we can look at access to clinical trials, access to palliative care, and resource allocation for life support. Each chapter outlines relevant core bioethics principles in more detail.

If I Were Restricted to One Film: *Wit* (2001)

Wit is likely the most well known film in this collection; and for good reason. Centering on a literature professor with end-stage ovarian cancer, this film is the quintessential teaching film in American medical schools regarding end of life, code status discussions, nursing ethics, moral distress and death and dying. The film also teaches what is known as the "hidden curriculum"—bedside behaviors and professionalism. However, for some students, it also serves as an introduction to narrative medicine (not just because of the obvious fact that it is a film about medicine), but by introducing the student to the poetry of John Donne (1572–1631). In fact, this film would be equally at home in a poetry or medical humanities course.

The Story Behind *Wit*

Wit was produced by the Home Box Office (HBO) network, which had transformed "made-for-television" films into using high-quality production standards that had typically been reserved for theatrical release, or first-run films. The cast and crew of *Wit* were all well established, having won critical acclaim for their bodies of work. Adapted from a Pulitzer Prize-winning stage play, the film is directed by Mike Nichols, with the screenplay co-authored by Nichols and Emma Thompson. The film won an Emmy award for Best Film.

Wit debuted on HBO March 24, 2001, and, ominously, was released on DVD on September 11, 2001. The timing of *Wit*'s wide release in a freshly traumatized country did have an impact on its audience viewing it in the aftermath of 9/11, which is discussed further on. However, the origins of the film took root during the 1990s, which was a socio-political timeframe considered to be the most peaceful and prosperous point of the twentieth century.

© Springer International Publishing AG, part of Springer Nature 2018
M. S. Rosenthal, *Clinical Ethics on Film*,
https://doi.org/10.1007/978-3-319-90374-3_1

"*Wit*": The Play

Wit is based on Margaret Edson's 1995 play by the same name, (spelled in some editions of the play as "W;t" where the semi-colon is stylized as the "i"). It won the Pulitzer Prize for Drama in 1999. Actress Kathleen Chalfant played the original lead in multiple stage productions of the play, while actress Judith Light played the lead in the Broadway version. This is the only play written by Margaret Edson, who since became a kindergarten teacher. The play is based on Edson's experiences in 1985 working as a clerk in an oncology/AIDS inpatient unit at a research hospital in Washington, D.C. in which patients were participating in clinical trials for AZT and ovarian cancer. (Some suspect Edson worked at the NIH, but Edson has never revealed the name of the research hospital, and gave fictional names to the drugs discussed in the play.) Edson was a graduate of Smith College, and majored in Renaissance history. After a year abroad, she accepted the job at the research hospital. She later worked at the St. Francis Center (now the Wendt Center for Loss and Healing), writing grant proposals, which may have informed some of the content for the play. Said Edson: "Because it was such a low level job, I was able to really see a lot of things first hand" (Lehrer 1999). Edson began writing the play during the summer of 1991 just before she was to begin classes at Georgetown University to pursue her MA in English.

The character of "E.M. Ashford", Dr. Bearing's Professor in graduate school, is based on Donne scholar, Dr. Helen Gardner (Cohen 2000), who was the editor of an authoritative edition of Donne's work: <u>John Donne: the Divine Poems</u> (1952), in which she restores the text to its original and intended versions of spelling and punctuation based on original works published in 1633 and 1635. Edson reached out to Gardner along with one of her mentors at Smith College for help with the Donne content. Gardner schools Edson on the importance of punctuation in poetry, and this is why a semi-colon replaces the "i" in the original published versions of the drama.

The play was sent to every theater in the country, and was finally accepted for development in 1995 by South Coast Repertory in Costa Mesa, California. The play was first performed at Longworth Theatre in 1997. It was written as a one-woman play initially. The script went through several edits, and ultimately, about an hour of content was cut before the final production script was performed. As with many major literary works, this gem required many years of polishing and rewrites for it to "shine" as the enduring drama it remains. The final script was informed by edits made by dramaturge Jerry Patch and original director Martin Benson. The original production won a number of Los Angeles Drama Critics Awards. From its inception to first performance, Edson had become a teacher, and concluded that "Wit" was her unique contribution. On the East Coast, "Wit" was first produced by Derek Anson Jones (a highschool friend of Edson's), who directed it at the Long Wharf Theatre in New Haven, Connecticut in October 1997. It then opened in 1998 at the MCC theater in New York City to rave reviews (Marks 1998). There was no Broadway production of this play until 2012—long after the film was released.

Adaptation for Film

When HBO expressed interest in adapting the play for film, the actresses who performed it on stage were interested in appearing in the film version, but there was consensus from HBO and Nichols that the film version required a more prominent actor with more gravitas on the large screen (Lyall 2001); Emma Thompson was selected for the role, who also helped to adapt the screenplay.

The producers and director not only sought to capitalize on Emma Thompson's star power, but also envisioned Thompson's success in a cinematic adaptation of the play that would be *different* from an onstage production. According to Collin Callender, the president of films at HBO and producer Cary Brokaw (Pressley 2000):

> If we had cast Kathleen Chalfant or Judith Light, it would have felt like filming the play,' Mr. Brokaw said. 'But we made a conscious decision to not do that, but to give it new life as a film while capturing all the qualities of the play we so admired and responded to.... Adapting the play enabled Mr. Nichols to give the material an intimacy – the viewer can easily imagine herself sitting beside Ms. Thompson's hospital bed – that wasn't possible on stage. 'The play was brilliantly done and technically very difficult, and the material lent itself particularly well to being a film,' Mr. Nichols said. 'The script is very interior. There are lots of places to go, both in the head and in the past and in the hospital.' Ms. Thompson agreed. 'The script is interpretable in lots of different ways, and on stage I'd have had to play it completely differently,' she said. 'On stage, you don't have the advantages of film: you don't have close-ups; you can't whisper and go very quiet.'

Nichols started as a director on Broadway, so was very familiar with the distinctions of a play and a film, having directed and won awards for several Neil Simon plays, including "Barefoot in the Park", "The Odd Couple" and "Plaza Suite". The first film he directed was Edward Albee's *Who's Afraid of Virginia Woolf?* (1966), which earned him an Academy Award nomination and Oscar win for Elizabeth Taylor as Best Actress. He won his Academy Award for his next film, *The Graduate* (1967). Nichols began his career working as part of a comedy team with Elaine May and Paul Sills, founding the comedy troupe, The Compass (later renamed Second City), the progenitor of *Saturday Night Live*. Nichols' body of work was particularly pronounced by strong women, who were often the focus of his films, which also include *Silkwood* (1983) and *Postcards from the Edge* (1990), the adaptation of Carrie Fisher's semi-autobiography.

Nichols' interest in medicine is notable; his father was a doctor, who immigrated to the U.S. when Nichols was seven years old, and changed the family surname from Peschkowsky to Nichols. Nichols is also co-founder with Cynthia O'Neal of the non-profit foundation, Friends in Deed, which provides emotional, spiritual and psychological support to people affected by life-threatening illnesses, such as cancer or AIDS. *Wit* is not Nichols first medical film; he also directed *Regarding Henry*, a film about a brain injury patient.

Emma Thompson, who plays Dr. Vivian Bearing in the film, had won numerous awards for her acting and writing, including her screenplay adaptation for *Sense and Sensibility* (1995). Nichols had previously worked with Thompson on *Primary Colors* (1998) and *Remains of the Day* (1993).

Synopsis

Both the stage play and film are essentially a one-woman performance entirely focused on one character: the patient. She is a poetry scholar, Professor (Dr.) Vivian Bearing, who is an expert in the poetry of John Donne. In a stark yet powerful opening scene lasting about five minutes, she is introduced to us as she receives the bad news that she has ovarian cancer from her oncologist and the Principal Investigator (PI) of her clinical trial, Dr. Harvey Kelekian (played by Christopher Lloyd); he is essentially a peer in a different discipline. A rushed truth-telling scene, followed by a rushed informed consent, is followed by a flash-forward to Dr. Bearing in her hospital bed throwing up, and admitting she should have asked more questions. From that beginning, we see in subsequent scenes both the progress of her disease and, in flashbacks, learn of her life before the diagnosis as a developing scholar and then, renowned expert and full Professor.

The film is about Dr. Bearing's end of life experience, having consented to a clinical trial as a last resort, but not fully appreciating its consequences on her quality of life. The film retains much of the one-woman focus in that it breaks the "fourth wall" as the character interacts directly with the audience. However, because of Dr. Bearing's scholarship and expertise in Donne's poems that focus on death, Bearing's memories of her own training, and own teaching about Donne's works, are eventually informing her own experiences about death and dying. She realizes, as her own mentor (fictional Professor E.M. Ashford) had expressed to her when she was a graduate student, that the way Donne's poems are punctuated in various editions actually *do* matter if one is to understand the truth about death: that death is a "comma" and not an exclamation mark. She instructs Bearing that Donne's original punctuation was changed in later editions, which have added dramatic capitalizations and exclamation points that were never in the original, nor what Donne intended. She is reminded: "Nothing but a breath, a comma, separates death from everlasting afterlife." And, as Donne writes: "And death shall be no more, Death, thou shalt die."

Dr. Bearing is also dying alone; she has no family or friends who surround her, and we are left to wonder why this is. Nevertheless, it is the hospital staff and faculty who are her sole contacts and supports as she goes through what can only be described as a "bad hospital death"—without proper pain management or palliative care, or even pastoral care. Her relationship with her nurse, Susie, becomes the most important relationship as she loses more control over what happens to her. Susie has a code status discussion with Dr. Bearing, who decides to be "Do Not Resuscitate" (DNR).

The Oncology Fellow, Jason Posner, an archetype we've all met, is a former student of Dr. Bearing's who works with Dr. Kelikian (the Attending physician and PI)., He is tasked with ensuring Dr. Bearing follows the clinical trial protocol. Jason, who is a science "wonk" with poor bedside skills, embodies what every cancer patient ultimately encounters in an academic medical center: a cancer researcher far more interested in the patient as research, than the patient as person.

Toward the end, we do see the human side of Jason (for better or worse) in the "code scene" in which Dr. Bearing's DNR status is ignored, her dignity and autonomy violated, as Jason panics and calls the code team (presumably because he wants to save her, and is more attached to the patient than he lets on). Chaos ensues as the code is botched, leaving Susie the nurse morally distressed and traumatized.

The Social Location of Wit

There are two social contexts to discuss: the context of the 1990s, when the play is conceived and developed, and the social context of the film's debut in 2001.

The Play: 1990s

Margaret Edson developed "Wit" as a play (completed in 1995) during the heyday of the Clinton Administration (1993–2000) before it was rocked by the White House intern scandal and impeachment proceedings (1998–99), in which President Clinton finally admitted to having had sexual contact with an intern, Monica Lewinsky. The pre-Lewinsky Clinton era was considered one of the most optimistic periods in U.S. history in a post-Cold War era, and *Time* magazine's cover after Clinton's election win read: "The Democrats' New Generation" referring to the first generation of baby boomers to access the highest levels of government. This period also represented great strides in home technologies, as Americans began to access the internet, enabled through a 1991 bill known as the *High Performance Computing and Communication Act* (which promoted the "information superhighway"), sponsored by then Senator Al Gore, who would become Vice President in 1992. Most Americans did not really have the necessary upgrades to access the internet much before 1995 when Microsoft bundled Internet Explorer and Windows '95 into a turnkey software package. In 1991, there were no commercial websites anyone could go to; by 1999, there were under 20 million websites, which jumped to just under 40 million by 2000. In 1995, Americans were just learning how to "surf the net" as the "Start Me Up" Windows '95 commercial introduced. With access to the World Wide Web, Americans entered a period of entrepreneurial prosperity during the "dot com bubble" (1997–2000). Many began to go online for their health information, in particular, as the early internet sites were dating sites and health sites.

Historic numbers of women started to go through perimenopause in the 1990s as the first generation of baby boomers (1946–1964) entered their 40s and 50s. As discussed further (See under Medical History Context), there were also significant numbers of women in the 1990s who had delayed childbirth for social or professional reasons, who were seeking out assisted conception with "artificial reproductive technologies" (ART) in which fertility work-ups and treatments remained unregulated (Rosenthal 1998, 2002).

The 1990s is also a timeframe in which the burgeoning field of feminist ethics (and especially, feminist bioethics), discussed further below, is in tension with a growing "feminist backlash" movement, a term coined by Susan Faludi in her 1991 book (Faludi 1991). Hints of this backlash emerged when Hillary Clinton began to reshape the role of First Lady (taking an office in the West Wing), having famously defended being a working mother on the 1992 campaign trail instead of staying home "to bake cookies" (Nicks 2015).

By now, women begin to realize that there is a myth of "having it all" and that social gains in the workplace are clashing with their biology, childcare responsibilities, and more conservative viewpoints (known as "anti-feminism") surrounding reproductive rights and gender preferences. On the small screen, *The Mary Tyler Moore Show* that shaped the 1970s is replaced by *Murphy Brown* (1989–1998), in which its heroine is continuously confused about her social role, and becomes a single parent—a plot line actually mocked at the 1992 Republican Convention by Vice President, Dan Quayle. *Ally McBeal* (1997–2002) takes women's social role confusion to new levels, as the series resonates with a younger generation of women who both flourish and suffer from "too many choices"; the character is also plagued by a repeating image of a "dancing baby" as her biological clock becomes stressful. Feminist scholars were aghast when a June 29, 1998 cover story of *Time* magazine put Ally McBeal on its cover alongside Susan B. Anthony, Betty Friedan and Gloria Steinem with the title: "Is Feminism Dead?" Indeed, this period marked a turning point in younger women rejecting "feminism". A flurry of popular articles and books emerged surrounding the choice to leave professional roles and embrace motherhood renamed to "stay-at-home-Moms". In reality, these were not so much "choices" (Wolf 2001; Rosenthal 2002) but a reflection of failed family-friendly policies that supported women in the workplace. The Clinton Administration's early cabinet picks made these problems painfully obvious; the first pick in 1993 for a female Attorney General nominee, Zoe Baird, for example, had to concede that she had hired a nanny who was an undocumented immigrant (known as "Nannygate"). Women were receiving confusing messages about feminism, on one hand, and pronatalism on the other. They began to resent the "double duty" or "second shift" roles. Terms such as "sandwich generation" also emerged to describe baby boomers caring for children and older parents at once, as statistics surrounding unpaid caregiving started to be acquired. The term "Glass Ceiling" (coined in 1978) is first used in a federal report title in 1995 (Federal Glass Ceiling Commission 1995), which acknowledged that women lost ground in their careers when they had children, dubbed "the Mommy track". This was extremely common in academia, as women became less productive on the tenure clock. Popular books women were reading around this time included What Our Mothers Didn't Tell Us (Crittenden 1999) and What's A Smart Woman Like You Doing at Home? (Burton et al. 1993). Feminist scholars at the time bemoaned unfortunate definitions of feminism as partly to blame for the "backlash". It was not that "women were equal to men" and therefore, needed no accommodations for their biological differences and social roles. More scholarly definitions of feminism demanded recognition of unfair social arrangements, and recognition that women's interests and men's interests *were of equal societal value*. Thus, workplaces that created family-friendly policies would benefit society as a whole—not just working women. But this didn't happen. Instead, women's careers thrived when they chose not to have children; the term "childfree" began to replace the term "childless" around this time.

Margaret Edson creates a protagonist within this social milieu: Vivian Bearing, Ph.D., is a full professor and expert in her field of seventeenth century poetry and the works of John Donne. She is a classic baby boomer who is 50; likely in perimenopause; single; and has made the "trade" of building a professional career instead of a family. A history of no pregnancies and her age puts her in the classic risk group for ovarian cancer. Dr. Bearing is the "poster patient" for the feminist backlash movement. She is a powerful intellectual woman who rose to the top of her profession who is now dying alone as her reproductive organs have rebelled.

"Wit" resonated on stage (and later on film) with single females, in particular, who were beginning to panic about being alone by choice or by chance. In 1993 (the year of the first World Trade Center bombing), the film *Sleepless in Seattle* repeated an eerie, untrue *Newsweek* statistic originally reported in 1986: "It's easier to be killed by a terrorist than it is to find a husband over the age of 40". The film makes the point that while the statistic was proven *untrue*, it still "feels true" to many women of that era (Garber 2016).

By the late 1990s, the HBO television series, *Sex and the City* (1998–2004) focused on single women. In one episode (1999), the character Amanda, a successful lawyer played by Cynthia Nixon (who later is in a revived Broadway production of *Wit*) orders takeout Chinese food and has a near-death experience as she chokes on her Kung Pao chicken. She exclaims to her best friend, Carrie Bradshaw: "Carrie – I'm going to die alone".

Two years later, the same year *Wit* debuts as a film, *Bridget Jones Diary* (2001) opens with Bridget fearing she will die alone in her apartment and not be found for days. *Wit* confronts a real problem in end of life care: people dying alone, and for women, in particular, the fear that they will die alone with no social support. In this timeframe, more women, for a variety of reasons, are without life partners, and are facing death by themselves.

The Film: 2001 and Post-9/11

Wit debuts in March 2001 during the early days of the George W. Bush Presidency —the result of a Supreme Court decision (*Bush v. Gore*) in which the courts weighed in on an historically tied election (2000) that was fraught with complications of a Florida recount. There was a sense by half the country that George W. Bush, who primarily ran a campaign focused on domestic issues and tax cuts, was not a legitimate president as a result of the Supreme Court's decision. Still, the country was coming out of a very prosperous and peaceful time with a large surplus with the exception of the "dot com" bust of 2000—in which several young internet companies failed early. Overall, the country was generally polarized around domestic policy and the Monica Lewinsky scandal. There was still a pervading sense of security, prosperity and optimism that dominated the turn of the millennium. Terrorism and national security were not topics discussed during the 2000 campaign season (Rosenthal 2013).

The cable viewing audience was becoming more fragmented in terms of demographics and marketing, and traditional movie theaters were no longer the venue of choice; more people (especially singles) were buying cable packages, and HBO was the first cable channel to develop not just independent high quality series, such as *Sex and the City*, but independent films of comparable high quality to first-run films for traditional movie theaters. People began to stay home for entertainment, while technology improved and provided more options. Social media had not yet been invented, but the early use of text messaging through Blackberry and newer model cell phones had begun.

Everything changed on September 11, 2001, which was the official release date of *Wit* on DVD. Viewers who saw *Wit* when it aired on HBO in March, had a different viewing contextual experience than viewers who watched it on DVD after September 12, 2001.

In a freshly post-9/11 world, during the aftermath period (9/12-01-9/11/02) the entire country was living through grief and going through a "mortality moment". As those who were adults on 9/11 recall, the attacks affected families coast-to-coast; end of life calls left on voicemail or dying callers on the phone with family or first responders, became the iconic sounds of 9/11 in the audio archive (Rosenthal 2013). Visual images of people jumping from the towers, as a "palliative" option became the iconic images of the visual archive (Rosenthal 2013). Death and dying, bereavement, funerals and memorials were now content for what used to be the cheery morning news shows that mainly stuck to weather and fuzzy human interest stories. On one CBS morning show that aired on September 12, 2001 for example (9/11 Television Archive 2001), grieving families (widows and children) whose loved ones worked for Cantor Fitzgerald on the impact floors of the North Tower, were interviewed in their fresh grief while viewers became vicariously traumatized themselves. *Larry King Live* regularly featured, throughout the 9/11 aftermath period, families with missing or dead loved ones (9/11 Television Archive 2001). The news coverage was divided into fast-moving policies surrounding national security that passed through Congress swiftly, to the "grief" coverage. In fact, the 9/11 Commission was initiated by grieving widows in New Jersey, who wanted to know why their husbands were killed, and how the country had failed them (Breitweiser 2006).

Amidst this complete shift in American history and American "innocence", the film version of *Wit* powerfully resonated, and emotionally devastated, audiences as they watched someone who "used to feel sure" (as Vivian Bearing relays to Susie) enter a death spiral in which everything feels out of control. *Wit* was ideal for home viewing due to its impact on viewers—particularly women—who found themselves shaken and unglued by the end. With its strong focus on the poetry of John Donne, and a country overwhelmed by images of death, the film, *Wit*, confronts both individual, and national, mortality issues that have still not been resolved in a country forever changed by 9/11.

History of Medicine Context

When teaching *Wit*, there are four main history of medicine contexts to cover: the state of cancer research from 1991 to 2001—the timeline involved from the playwright's conception to the adaptation to film; the state of women's health 1991–2001, and particularly, gynecologic cancer risk factors and treatments, including effects of fertility drug treatment, which was unregulated, and in response to the social production of infertility from delaying conception (see above). One must also review the state of end of life care during this timeframe.

There is fourth history of medicine context some may find worth teaching: the context of John Donne's works in plague-ridden London during his lifetime (1572–1633). Donne's most famous line: "never send to know for whom the bell tolls, it tolls for thee" in his work, Devotions Upon Emergent Occasions and Severall Steps in my Sicknes (1624) referred to bells that were rung each time there was a new death. If *Wit* is shown in a medical school class, one may choose to time infectious disease sections around this content, for example. Cancer was often referred to as the "plague" of the twentieth century as cancer deaths were rising, and predicted to overwhelm healthcare systems. Below, each of these contexts is reviewed separately. As a general overview for what is taking place in the 1990s during this timeframe, the biggest debate emerges over access to healthcare when the Clinton Administration attempts to pass the *Healthcare Security Act*, which dies its own death in Congress (discussed in Healthcare Ethics on Film). Other major breakthroughs were in cloning and the establishment and completion of the Human Genome Project, which is touched on more further on.

Cancer Research in the 1990s

To feature a female character in a cancer clinical trial in the 1990s was a sign of a recent sea change in clinical research. In fact, more women were recruited into clinical trials during this period than in any previous time in U.S. history. This was due to a new recognition that women had been traditionally excluded from clinical trials, and there was an effort to enroll them so that the burdens and benefits of research could be distributed more equally between the sexes, while meaningful data about gender differences in drug efficacy and dosing could be collected. In 1990, an extraordinary thing occurred: the Office of Research on Women's Health was established within the Office of the Director at the NIH. This was based on the influence of feminist bioethicists as a field, which I discuss further on.

In terms of therapeutic gains, this was an especially productive decade in cancer research, in which there was some evidence that overall deaths from cancer were declining while more meaningful clinical trials were being developed (Scheibinger 2003). By this point, the FDA had created the Investigational New Drug (IND) trials, in which patients could receive an investigational drug outside the normal "blinded" research setting (White Junod 2008). Between 1990 and 1992 FDA guidelines established a "parallel track approval" process to make special categories of drugs available to wider group of patients. The FDA also accelerated approval of new drugs based on "surrogate markers" (a.k.a. tumor markers) to widen therapeutic possibilities, which also contributed to therapeutic misconceptions (White Junod 2008). (By 2014, studies showed that very few of such cancer drugs wound up actually extended life or improved quality of life.)

In 1994, FDA made changes in its policies to facilitate women's participation in the earliest phases of clinical drug trials in light of the push at the NIH for the inclusion of women in clinical trials (see further).

Between 1991 and 1995, the National Cancer Institute (NCI), Centers for Disease Control (CDC) and the American Cancer Society (ACS) reported that cancer deaths had dropped by 2.6%; by 2008, cancer deaths dropped further to 18%, which was mostly due to improved treatments and screening (ASCO 2017). Primary prevention efforts were also on the rise. Major breakthroughs in the 1990s also included anti-nausea drugs, such as Zofran (1991), which allowed patients to tolerate chemotherapy (and drug trials) more easily, and often in an outpatient setting (ASCO 2017). Edson is working at a cancer research unit during this timeframe.

State of Women's Health: 1991–2001

Up until this timeframe, there had been a cardinal rule in clinical research that women, in general, should be protected from being recruited for research because they were a vulnerable population. Past problems with thalidomide, for example, and unethical trials with the Pill in Puerto Rico, created concerns that potential risks of research on women subjects could outweigh benefits due to unknown risks, especially to women of childbearing age, or who were pregnant.

However, it also became clear that the *exclusion* of women in research was becoming problematic. The most notable example was in data surrounding the protective features of aspirin in cardiovascular health (Scheibinger 2003; Mastroianni et al. 1994). It was falsely posited that men were more at risk for cardiovascular disease than women, when this was not the case. It was only that cardiovascular symptoms and presentation were *different* in men than women. In a variety of settings, feminist bioethicists argued that the medical community was making false assumptions about targeted therapies for women, based only on data from male subjects. It was also making false assumptions about the presentation of disease symptoms in women based only on data collected from men. Additional arguments were that women were not able to access the potential benefits of research, nor were women's health issues being considered as worthy of researching as men's. One of the purposes of the Office of Women Health Research was to address "inequities in federally funded biomedical research, diagnosis, and treatment of diseases affecting women...".

The preface to a 1994 Institute of Medicine report stated:

A perception has grown in recent years that biomedical research has focused more o the health problems of men than on those of women, and that women have been denied access to advances in medical diagnosis and therapy as a result of being excluded from clinical studies. The *NIH Revitalization Act* of June 1993 introduced new requirements for the inclusion of women and minorities in federally funded clinical studies except where their exclusion [is justified]. Congress intended the Act to be responsive to the serious allegations of 'underrepresentation of women in clinical reearc, and to be corrective of any disaparities.... The inclusion of women in clinical studies raises a mélange of often conflicting ethical, legal, scientific and social traditions and concerns. (Mastroianni et al. 1994)

The Infertility Epidemic and Ovarian Cancer Risk

To bring the protagonist's illness back into focus, it's important to discuss increases in ovarian cancer incidence at this time due to iatrogenic causes. Alongside concerns that women were under-represented in research, they were over-represented in high-risk fertility treatments in which women subjected themselves to a host of experimental or innovative therapies with low success rates. In the 1990s, there was indeed an infertility "epidemic" that was largely socially produced, due to women delaying childbirth, which affected fecundity rates (see under Social location). There were also increases in pelvic inflammatory disease due to unprotected sex and HPV, which caused blocked fallopian tubes. Other causes for infertility burgeoning at that time were concerns about estrogen mimics from organochlorines, which also affected male fertility.

Feeling pressure to have children, women who struggled with fertility sought out a range of assisted conception techniques, which, at its core, required their cycles to be controlled to artificially induce ovulation. By the late 1990s (Glud et al. 1998), it became clear that these treatments put women at higher risk of ovarian cancer, but more recent analyses (Diergaard and Kurta 2014) point out that women at highest risk were those who sought fertility treatment who never conceived; or those who may have already had the BRCA1 or BRCA2 mutations.

Fertility treatment becomes a dominant area in bioethics debates at this time because of concerns over unknown harms, and its potential abuses for sex selection and trait selection. The discovery of cloning in 1996, which made science news featuring Dolly the Sheep (cloned using a mammary cell) only magnifies concerns as there are fears that cloning may become an "option" that is abused surrounding fertility treatments. With respect to the BRCA discoveries, debates continue to rage over the use of genetic testing and unintended privacy concerns; this, too, becomes magnified with the announcement of the Human Genome Project in 1997, which is completed early in 2000—all during this timeframe. The film *Gattaca* (1997), a well-known bioethics film, reacts to social anxieties about cloning and the Human Genome Project. It depicts a dystopian future in which access to social goods are based solely on genetic profiling, which makes "natural conception" no longer the norm.

Gynecological Cancer Treatment

Although advances had been made in breast and cervical cancer rates due to better screening programs (Pap smears and mammography), there was no screening test for ovarian cancer, which in most cases involved epithelial cell tumors that covered the surface of the ovaries. By 2000, ovarian cancer treatment remained grim; only 25% of ovarian cancers in the U.S. were diagnosed at the early stages (Cohen 2000). By around 2005, a still-unreliable blood test, a tumor marker known as CA-125, was being used to help detect ovarian cancer (Moss et al. 2005; Dorigo 2011), but its reliability was questionable.

Thus, ovarian cancer was very much in the news and on the minds of women during this timeframe because it was hard to diagnose and there were no good treatments for advanced stages.

Ordinary risk absent of any fertility treatments were noted in women who were never pregnant or breastfed. Single women were advised to use oral contraception, if possible, so they could gain the unintended benefit of reducing ovarian cancer risk (Modan et al. 2001; Rosenthal 2002).

Advances in ovarian cancer treatments were made around 1992–4, when a new family of drugs emerged through paclitaxel (Taxol). These drugs, when combined with cisplatin, were shown to shrink ovarian tumors more than half in patients who did not respond to other therapies; it also had extended survival by about a year (ASCO 2017). Eventually, paclitaxel is also used in breast cancer as an adjuvant therapy. This cancer drug was derived from the bark of a yew tree, and was among several cancer drugs originating from natural sources. Until drug makers discovered a synthetic method for producing it, there was concern the natural resources needed to produce the drug would not meet demand (ASCO 2017). These concerns were depicted in a popular film of that time, *Medicine Man* (1992), in which a cancer drug is discovered within a threatened rain forest.

A milestone in women's cancer research occurs between 1990 and 1994 when the discovery of the BRCA1 and BRCA2 gene mutations that cause breast and ovarian cancer are published (Park 2014). For the first time, prophylactic treatments are developed for women at risk, and genetic testing emerges as a significant part of cancer prevention and treatment. In women with the mutations, prophylactic therapies were found to reduce their cancer risk by 90 percent with prophylactic mastectomies and oophorectomies.

More advances are made in 1998 when Tamoxifen, an estrogen-blocking drug, is used in reducing breast cancer recurrences in women with estrogen-receptive tumors. That same year (1998), trials with the drug, Herceptin, which was more targeted therapy, and less toxic to other cells, began around 1998. Neoadjuvant chemotherapy is also introduced, which involved chemotherapy prior to surgery. Such surgeries helped to shrink large breast tumors to allow more women to undergo lumpectomy instead of full mastectomy (ASCO 2017).

End of Life Care in the 1990s

End of life care in the 1990s was not well addressed because of general resistance by practitioners to palliative care. Aggressive therapies for end-stage cancers were pursued for two main reasons: first, there was tremendous interest by patients to enter cancer drug trials in order to receive a chance at benefiting from newer drugs;

in the U.S., such trials also provided free care to patients regardless of their insurance coverage. Second, practitioners in this period were trained to view aggressive care as "beneficent" and "state of the art," and palliative care as "giving up" or abandoning patients.

Advance directives were just beginning to become mainstream, too. In the Nancy Cruzan case (Cruzan v. Director 1990), discussed more in Chap. 2, a case that involved the request to stop nutrition and hydration for a patient in a persistent vegetative state (PVS), the court recognized the constitutional right of competent persons—or their surrogates—to direct their own care at the end of life, which included refusing life-sustaining therapy. Patients could also elect to decline cardiopulmonary resuscitation (CPR) by indicating their code status preference as DNR. The creation of Advance Directives was the outcome of the Cruzan case, which led to the *Patient Self-Determination Act* (1990) in which patients could make their wishes known about a variety of treatments, including who they wished to be their surrogate decision-makers. This law newly required hospitals to provide written information about Advance Directives to all patients.

Nonetheless, the vast majority of patients during this time frame faced uphill battles to de-escalate care or receive adequate pain control. Palliative care in the 1990s was not new or novel; but few hospitals had trained palliative care professionals or services, and it would take another 20 years before the vast majority of academic medical centers recognized the need for comprehensive on-site palliative care services, partly due to the recognition that aggressive care did not necessarily improve quality of life or outcomes (Connors et al. 1995). It gained steady growth since 2000, but even as late as 2011, 50% of U.S. states did not have optimal access to palliative care services (CAPC 2015).

The Medical Context of John Donne's Lifespan (1572–1631)

Edson explicitly brings the works of John Donne into the drama by making his seventeenth century words relevant to a modern death experience. Although the audience learns about Donne's works through flashpoints in Dr. Bearing's memory with respect to her early understanding, later teaching, and current epiphanies, about Donne, any teaching of *Wit* would be enhanced by a more comprehensive summary of the medical context of Donne's works: the repetitive plagues of London. This can be juxtaposed with Edson's own experiences of witnessing AIDS and cancer research trials, as both have often been referred to as our modern "plagues". Donne's Holy Sonnets (thought to have been inspired by his grief over his wife's death) were published two years after his own death in 1633. But they were written *prior* to a work published while he was alive: <u>Devotions Upon Emergent Occasions and Severall Steps in my Sickness</u>. It is this work that is considered to be his most enduring narrative medicine work, which yielded one of the most famous "death and dying" stanzas still cited frequently:

No man is an island, entire of itself; every man is a piece of the continent, a part of the main. If a clod be washed away by the sea, Europe is the less, as well as if a promontory were, as well as if a manor of thy friend's or of thine own were: any man's death diminishes me, because I am involved in mankind, and therefore never send to know for whom the bells tolls; it tolls for thee. (Donne 1624 "Meditation XVII")

Devotions was written while Donne was bedridden, feverish and delirious at times when he succumbed to typhus during the "spotted fever" epidemic of London in 1623–24. He became ill in November 1623, and recovered in late December. Through his illness, he wrote Devotions, which comprises 23 sections, which scholars have divided into "Meditations", "Expostulations", and "Prayer". His famous "Meditation XVII" (Meditation 17) is the most quoted. Donne's musings about the course of his own illness in Devotions mirrors what some scholars note are comparable to medical chart notes that would have been made in the morning, afternoon and evenings (Lander 1971). Devotions was entered in the Stationers' Register on January 9, 1624 (Lander 1971) and published shortly thereafter.

Donne's entire lifespan knew nothing but bouts of plagues, which shaped the social and economic rhythms of this world. Since the first appearance of "Black Death" (bubonic plague) in 1348, until London's final outbreak of the disease in 1665, plague outbreaks would strike at least 40 times; there were major outbreaks about every 20 years, killing 20% of the population, but there were continuous lesser outbreaks in between. Bubonic plague was caused by the bacteria, *Yersinia pestis*; it was transmitted when a human was bitten by an infected flea. The fleas were carried into many urban centers by the common black rat and other rodents, who also died from the bacteria; the fleas would jump from the dead rodent to humans to find new hosts. In general, humans contracted bubonic plague from both flea bites and other infected people through their bodily fluids, including cough spewings. The worst bout of plague was in 1563, when 24% of London's population died a decade before Donne was born. The plague outbreak of 1665 killed 100,000 people within seven months. Generally, 60–80% of those infected died within a week, as the plague quickly attacked the lymphatic system and the lungs.

"Bills of Mortality" began to be published regularly in 1603, in which year 33,347 deaths were recorded from plague. In Donne's lifetime, there was virtually no year when there were no recorded cases of plague. In 1623–24, a "spotted fever" epidemic hit, which was not specified as typhus (Lander 1971). In 1625, there were about 41,313 persons recorded dead, which became known the 'Great Plague', until 1665 surpassed it, after Donne's death (Museum of London 2017).

In the 1600s, "plague bells" were rung at burials for 45 min as warnings that precautions needed to be practiced, which was basically social distancing; these bells are constantly mentioned in Devotions, including "for whom the bell tolls". Donne even writes about such precautions against the 1625 plague himself: "the infection struck into the town into so many houses, as that it became ill manners to make any visits." (Lander 1971) Donne seems to have avoided contracting the bubonic plague, but he did not avoid contracting typhus, which had raged during 1623–25, and overlapped with the Great Plague year. At the time, any other illness other than the plague was considered "minor", but typhus fever was still very serious, and historians believe it may have been confused with plague, contributing to the high mortalities of 1625. Typhus was also carried by rat and flea infestations in crowded centers. Writes Donne scholar, Lander (1971): "The gutters of Donne's London were alive with rats, the cobbles littered with refuse, and cesspools connected with latrines were emptied only after they overflowed. Britain itself had long since ceased to be an island and had become instead a kind of suburban cesspool of the mainland."

Layered onto bouts of plague were typhus outbreaks that were both endemic and epidemic (Lander 1971). The book, Devotions essentially follows the typical pattern of typhus; Donne starts out by discussing his spots and rash, and then continues to write about his fevers, delirium, photophobia, tinnitus, etc. Donne also references how London had "depopulated" referring to continuous population migration out of urban centers to rural centers and vice versa, in an effort to escape outbreak situations.

Throughout Donne's works, "his essential philosophy [is that] suffering humanizes man. But his statement in Meditation I (p. 1, l. 26) is more succinct and even more macabre, 'we die, and cannot enjoy death' ... No more fitting epitaph could be accorded that quarter of the seventeenth century" (Lander 1971).

It is not known what, exactly, the cause of Donne's death is in 1631; some speculate it may have been a relapse of typhus, or perhaps a complication of it (enteritis). While writing Devotions, Donne was Dean of St. Paul's Cathedral in London.

In some ways, Dr. Bearing's character is following Donne's model in Devotions, by tracking her illness day by day using the "fourth wall". It is unclear whether Edson was familiar with Devotions, but it would be remiss not to discuss Devotions alongside Wit.

Clinical Ethics Themes

Clinical ethics themes in Wit comprise autonomy issues such as informed consent, truthful prognostication, Advance Directives and code status; and surrogacy. Other themes revolve around what constitutes a "beneficent" care plan (one that maximizes benefit and minimizes harm): the risks versus benefits of aggressive care within a clinical trial context versus palliative care or hospice for this patient. Nursing ethics and moral distress are also strong themes in this film, as are professionalism and humanism. While these themes are explored further below, it's important to note that whichever theme is the focus of a discussion—ultimately—we are forced to admit that Dr. Bearing has had a "bad death". Full stop. Particularly in a medical school environment, any discussion of the clinical ethics themes should force contemplation of what we could change to make Dr. Bearing's death a "better death" or, even, a "good death" experience. Thus, regardless of which themes or angles one focuses on below, the flawed consent process—particularly for enrollment in a clinical trial—combined with the absence of adequate institutional and social supports for this patient, should dominate the discussions.

Autonomy Issues

The first five minutes of *Wit* is a combination of a truth-telling and informed consent scene. The first three words of the film are: "You have cancer." Some would argue that the frank delivery is a serious faux pas; however, "best practices" such as the SPIKES model (Bailea et al. 2000) guide practitioners into tailoring news delivery to the patient's preference, and to first ask patients how they would like the practitioner to break down the news s/he has to deliver. Thus, we don't know whether Dr. Kelekian, in getting to know his patient before we enter the film, was encouraged to not "mince words" and just tell it like it is. We may presume that as a literary scholar, this patient would not appreciate unnecessary preamble. The SPIKES model, for example, emphasizes the importance of ensuring these discussions take place in a private setting (to the extent possible) and clearly stating "I have bad news" to give patients a moment to brace themselves emotionally. It remains unclear whether something of this nature was said prior to "You have cancer." There are more problems with the discussion of the treatment options. There are three components to informed consent: Disclosure, Decision-Making Capacity, and Voluntariness. In a good consent discussion, full disclosure of the various treatment options, their potential risks and benefits, are essential. This was not done, as one obvious option was palliative care, which was never discussed. Second, patients should be assessed for their capacity to make a decision; this is a medical determination the healthcare provider of record can make using some sort of "teach back" method, that invites the patient to summarize what she understands, and asks questions to determine whether she has enough understanding and appreciation to make a decision, and what questions she may have. Patients who say they have "no questions" after being given bad news indicates they may be overwhelmed, and practitioners may further suggest that patients take some time with the information and call with questions later. None of this was done. Instead, a consent form outlining a clinical trial, is shoved in front of her by Dr. Kelekian, who is also the Principal Investigator (PI) of the study himself. How would this affect his ability to be objective and discuss risk? In some ways there was definitely a "therapeutic misconception" on Dr. Bearing's part, that is, an exaggerated belief in the efficacy of the treatment being offered. What part of that misconception can be blamed on the informed consent process? In addition to "therapeutic misconception," there are other issues about Dr. Kelekian's research that may have influenced the medical options offered to patients, such as conflicts of interest. For example, if he had a financial interest in the research, how might this affect valid consent? Overt conflicts of interest exist when a PI is receiving direct compensation from industry; however, even with federally funded grants, promotion and tenure are based on funded grant activity and publications, while the validity of research is often dependent on the number of patients enrolled in the study, which may be another indirect conflict of interest. For example, what if Dr. Kelekian's need to increase his numbers of research subjects was one of the reasons he seemed to rush the informed consent process with Dr. Bearing? Morever, clinicians may be further ethically compromised when they are paid a fee for each patient they enroll in a clinical trial, more common in private practice. Notwithstanding, Dr. Kelekian seems more interested in selling enrollment to the patient than in walking her through other options, such as palliative care or hospice in the absence of a known, definitive treatment. Instead of voluntary or valid consent, Dr. Bearing hastily signs the form without reading it; denies there is anyone in her social circle that needs to be contacted, only to realize in a few weeks that the side-effects are unbearable, and that she "should have asked more questions." Thus, while there is, at least, truthful prognostication that could have used some stylistic finesse, there is a botched informed consent process that violates patient autonomy. In fact, one good question to raise in a class setting is whether Dr. Bearing's education *disserved* her because there was a presumption that she would understand complex oncology protocols (without any training in that topic). Had she had an average 10th grade education, as many patients do, there may have been much more effort involved in explaining treatment options. Dr. Kelekian notably states during the "consent" discussion: "The important thing is to take the full dose of chemotherapy." However, this is only "important" for the purposes of the research protocol; not necessarily for a beneficent treatment plan, as the full dose may have more risks than benefits for the patient. There is a conflict of interest/commitment when the PI is also in the role of healthcare provider, and the film accurately portrays such conflicts.

Surrogate Decision-Making

During the consent scene, Dr. Kelekian asks his patient if there is someone else who should be informed about her illness, and she flatly says, "No." This was mismanaged by the practitioner; competent patients with decision-making capacity who are about to undergo aggressive treatment may lose capacity as a result of the progression of their illness, such as pain, or side-effects of treatments. Surrogate decision-makers need to be identified in advance. It becomes evident in the film that part of the "bad death" we are witnessing is that Dr. Bearing is dying alone. In such cases, we need to identify a surrogate decision maker for her who will represent her wishes when she is no longer conscious or capacitated. Thus, is Dr. Bearing, in some respects, an "unbefriended" or "unrepresented" patient? The legislation on surrogate decision-making varies in different countries, states and provinces. Regardless of whether legislation bars practitioners from making substituted judgments for patients because they are deemed to have an inherent conflict of interest as a healthcare provider, or allows substituted judgments in the absence of any surrogate (presumably based on previously stated preferences or best interests), the ethically ideal scenario is for patients to identify an authentic surrogate decision-maker early. That way, should they lose capacity due to their illness, the Principle of Respect for Persons, which obligates practitioners to honor autonomous patient's wishes and protect non-autonomous patients (including those who lose capacity) is fulfilled. In this film, none of this is done, and the patient's own stated preferences surrounding code status are ignored.

Many audiences find E.M. Ashford's visit especially difficult to watch because it's clear that Ashford could have served as a potential support or surrogate for the patient. In this scene, the long retired Ashford pays a visit initially to Dr. Bearing's office, and is told she will find her in the hospital. Ashford explains that she was in town visiting her great-grandson, and we presume, she probably always takes the time when in town to see her former star pupil. When she offers to recite some of Donne at the bedside (perhaps even "Meditation XVII" from Devotions), Dr. Bearing groans "Nooooooo," and instead, Ashford pulls out a children's book—no doubt intended for her grandson—and starts to read The Runaway Bunny by Margaret Wise Brown (1942). Two renowned Donne scholars at a deathbed find the most comfort in a classic children's book—an unusual twist by Edson, who probably uses the book all the time as a Kindergarten teacher. If screening this in a classroom, many will begin to weep at this scene because we bemoan the lost opportunity for Dr. Bearing to have had a support system, and probably an out-standing surrogate decision-maker in Ashford. Undoubtedly, Ashford would have agreed to be the surrogate, and would have understood the process of death (a comma) better than the care team. Ashford's compassion for her dying former student is also testament to how proud she as of Dr. Bearing's accomplishments. She ends the visit by saying: "Time to go," and a line from "Hamlet": "And flights of angels sing thee to thy rest".

Advance Directives

This film illustrates what happens when an Advance Directive is violated and a code is mismanaged. Although the patient has clearly stated she wishes to be DNR should her heart stop, and the DNR order is entered by her Attending physician, none of this seems to matter. When she codes, she is treated as "full code". The mismanagement arises out of mostly organizational ethics problems. The young Fellow (Dr. Jason Posner) panics when he realizes his patient—his research subject —has coded; he even yells: "She's research!" It's clear he may never have per-sonally experienced a patient's death before. He immediately begins manual CPR despite her code status and calls a Code Blue. The Code Team does not check the patient's chart upon entering the room and begins an aggressive code. The nurse, Susie Monoham, tries to intervene to no avail; due to still-existing organizational hierarchies in which nurses are not taken seriously or valued, the nurse's protes-tations that this patient is "DNR" or "no code" is ignored. Chaos ensues, and the Fellow finally stops the code by admitting "I made a mistake". But the damage is done—and the damage extends far beyond dignity harms to the unconscious (and expired) patient. The nurse is morally harmed by this botched code—perhaps more than the patient at this point (see further under *Nursing Ethics and Moral Distress*). This scene is a good illustration of a nurse experiencing moral distress, discussed further.

Professionalism and Humanism

Wit illustrates different models of professional behaviors. Professionalism refers to appropriate professional behavior, while humanism refers to a "world view" of a healthcare professional that truly cares about his/her patients. Acting like you care is not the same thing as *actually* caring. There is a tension between a paternalistic and collegial approach by Dr. Kelekian. He seems to treat Dr. Bearing as an academic peer (which she is), and thereby dispenses with the usual "extras" that may supplement disclosure or consent discussions. It seems he views her as the perfect research subject since she respects scholarship as much as he does.

We also see the perils of putting patients in the hands of young, inexperienced residents or Fellows, who lack enough life experience to properly relate to their patients and patient suffering. For example, Dr. Bearing is forced to undergo a detailed medical history and pelvic exam by Dr. Posner, who mishandles it from every conceivable angle, ranging from bedside body language to overuse of technical jargon, to poor technique in a pelvic exam, running at the last minute to get "a girl" in the room for the exam (Nurse Monaham). The use of jargon as a way to keep patients at a distance is a recurring theme in the film, which Dr. Bearing notes. The patient is further dehumanized during rounds (she calls it "Grand Rounds"), which may either be done deliberately or is a script error. If the former, it's useful to note that patients may confuse "rounding" with "Grand Rounds", a term referring to a formal lecture in a hospital. Or, Dr. Bearing may be making a pun. If the former, it is a minor inaccuracy in the script worth pointing out during teaching.

Humanistic behaviors are modeled by the nurse, who embodies classic "caring versus curing" themes of nursing ethics, discussed further. However, in the most connected doctor-patient interaction we see with the Fellow, it is because the patient has reached out and asked him about his research, which he shares with her, and they bond as scholars.

Nursing Ethics and Moral Distress

If used in a nursing curriculum, *Wit* is ideal as a nursing ethics film, as it is the nurse who is essentially the primary caregiver, and who develops a bond with the patient. The nurse also experiences profound moral distress, and we presume, moral residue, as a result of the botched code, in which the patient's wishes are ignored despite the nurse's efforts to correct it. Moral distress is defined as a situation where a healthcare practitioner recognizes a moral problem but is constrained from acting on it due to either internal or external constraints. Constraints may be legal, or organizational. In the Code Scene, the nurse recognizes the violation of the patient's wishes, and repetitively tries to verbalize the problem, yet is ignored because she has no "voice" that within the organizational hierarchy. After the code team finally vacates when the error is acknowledged, Susie Monahan begins to cover up Vivian's exposed body. When Jason approaches the bed, Susie lifts up a hand in a silent gesture that demands that he back away. It is a gesture that communicates: "I NEED SOME SPACE; I HAVE BEEN HARMED". When nursing audiences see this for the first time, it is when many of them break down and cry out of recognition, as this is a collective experience.

We see signs of the nurse being ignored earlier, when she makes the recommendation to Dr. Kelekian in an earlier scene for patient-controlled pain medication, and is also promptly ignored when Dr. Kelekian overrules her request by ordering a morphine drip. Moral distress frequently occurs when nurses clearly disagree with an order or a care plan but have no moral authority to make a change. In teaching, I would suggest that a clinical ethics consultation is appropriate in these situations.

We see her more humanistic approach to caring when she applies hand cream to the patient once she becomes unconscious, and her effort to talk to the unconscious patient so that there is more dignity, and an attempt to acknowledge she is a person and not just a "body".

Indeed, it is the nurse who finally has a goals-of-care discussion, and raises the code status question. We may presume, based on this discussion (which occurs while they eat popsicles together), that the nurse has "moral residue" (Rosenthal et al. 2015) from past cases in which patients were not provided with code status options. This time, she is going to ensure that this discussion takes place. Unfortunately, the nurse learns that even when she takes careful steps to ensure that inappropriate codes are not performed, it doesn't work. This code scene discussion emphasizes, from Dr. Bearing's question—"Susie – you're still going to take care of me?"—that DNR does *not* mean that we cease to provide care; merely that it spares patients from the usual poor outcomes of CPR and advanced cardiovascular life support (ACLS). Some may prefer to use the term, DNAR (Do Not Attempt to Resuscitate) when reviewing this scene. Patients such as Dr. Bearing, with severe underlying illnesses (e.g., metastatic cancer, sepsis, pneumonia, or stroke) who suffer cardiopulmonary arrest have less than a 1% chance for long-term survival. Furthermore, the few who do survive CPR are likely to suffer brain damage, and even those who don't will most often request to never be resuscitated again (Bailea et al. 2000). It is not unusual for nurses to have code-status discussions, although ideally, it would be best if the Attending physicians had these discussions. Nurses often initiate such discussions if they have moral distress over its avoidance. In this case, even when code status is discussed at an optimal time, and entered in the chart, the nurse finds herself, once again, morally "blindsided". We wonder how many years Susie Monahan will be haunted by this situation. It may be useful to show some clips of nurses discussing moral distress from *The Moral Distress Education Project*, a self-guided web documentary available at www.moraldistressproject.org (Rosenthal et al. 2015). One interview recalls a similar "botched code" that makes *Wit* all the more realistic in its portrayal of moral distress amongst nurses.

Conclusions

Teaching *Wit* is an exercise in making "cutting room floor" decisions about the depth and breadth of topics, but if one needed to carve only one film into a medical curriculum, this would be the film to select. It can be shown using a multidisciplinary panel approach as well, with narrative medicine, bioethics, palliative care, and oncologists on the panel.

Film Stats and Trivia

- Christopher Lloyd (Dr. Harvey Kelekian), made his feature film debut in the Academy Award-winning *One Flew Over the Cuckoo's Nest* (1975), discussed in Chap. 6.
- Audra McDonald (Nurse Susie Monahan) is a Tony Award winner and opera singer.
- Playwright Harold Pinter makes a cameo appearance as the patient's father, Mr. Bearing in the film.
- The film's haunting classical score alternates between Dmitri Shostakovich's *String Quartet No. 15* (1974), used in the opening scene and Arvo Part's *Spiegel im Spiegel* (1978) (Cizmic 2006).
- Won Outstanding Directing for a MiniSeries, Movie or Special.
- Won Outstanding Made for Television Movie.

From Theatrical Poster

Based on: Wit by Margaret Edson
Screenplay: Emma Thompson, Mike Nichols
Director: Mike Nichols
Starring: Emma Thompson, Christopher Lloyd, Audra McDonald, Eileen Atkins
Theme music composer: Henryk Mikolaj Gorecki
Producer: Simon Bosanquet
Cinematography: Seamus McGarvey
Editor: John Bloom
Running time: 98 min
Production Company: Avenue Picture Productions
Distributor: HBO Films
Premiere Date (aired): March 24, 2001.
DVD (HBO Home in 16:9 format: September 11, 2001.

References

American Society of Clinical Oncology (ASCO). 2017. Cancer progress timeline. http://www. cancerprogress.net/timeline/major-milestones-against-cancer. Accessed 19 July 2017.

Bailea, W.F., R. Buckman, R. Lenzia, G. Globera, E.A. Bealea, and A.P. Kudelkab. 2000. SPIKES —a six-step protocol for delivering bad news: Application to the patient with cancer. *The Oncologist* 5: 302–311. http://theoncologist.alphamedpress.org/content/5/4/302.long.

Breitweiser, Kristen. 2006. *Wake-Up Call: The Political Education of a 9/11 Widow.* New York: Warner Books.

Burton, Linda, Janet Dittmer, and Cheri Loveless. 1993. *What's a Smart Woman Like You Doing at Home?* Merrifield, Virginia: Family&Home Network.

Center to Advance Palliative Care (CAPC). 2015. Palliative Care Report Card. http://reportcard.capc.org/pdf/state-by-state-report-card.pdf. Accessed 19 July 2017.

Cizmic, Maria. 2006. Of bodies and narratives: Musical representations of pain and illness in HBO's W;t. In *Sounding Off: Theorizing Disability in Music*, ed. Neil William Lerner and Joseph Nathan Strauss, 23–40. New York: Routledge.

Cohen, Carol. 2000. Margaret Edson's Wit—Audience Guide. Madison Repertory Theater. http://faculty.smu.edu/tmayo/witguide.htm. Accessed 19 July 2017.

Connors, A.F., N.V. Dawson, N.A. Desbiens, W.J. Fulkerson, L. Goldman, W.A. Knaus, J. Lynn, R.K. Oye, F.E. Harrell, R.S. Phillips, J. Teno, and N.S. Wenger. 1995. A controlled trial to improve care for seriously iii hospitalized patients: The Study to Understand Prognoses and Preferences for Outcomes and Risks of Treatments (SUPPORT). *Journal of the American Medical Association* 274:1591–1598. https://jamanetwork.com/journals/jama/article-abstract/391724?redirect=true.

Crittendon, Danielle. 1999. *What Our Mothers Didn't Tell Us: Why Happiness Eludes the Modern Woman*. New York: Simon and Schuster.

Cruzan v. Director, Missouri Department of Health, 497 U.S. 261 (1990).

Diergaard, B., and M.L. Kurta. 2014. Use of fertility drugs and risk of ovarian cancer. *Current Opinion in Obstetrics and Gynecology* 26:125–129. https://www.ncbi.nlm.nih.gov/pmc/articles/PMC4217689/.

Donne, John. 1624. *Devotions Upon Emergent Occasions and Severall Steps in my Sicknes*. London: A.M. for Thomas Iones (Note: Ebook version by The Project Gutenberg posted to: http://www.gutenberg.org/files/23772/23772-h/23772-h.htm).

Dorigo, O., and J.S. Berek. 2011. Personalizing CA125 levels for ovarian cancer screening. *Cancer Preventive Research* 4: 1356–9. http://cancerpreventionresearch.aacrjournals.org/content/4/9/1356.long.

Faludi, Susan. 1991. *Backlash: The Undeclared War Against American Women*. New York: Crown Publishing.

Garber, Megan. 2016. When Newsweek struck terror in the hearts of women. *The Atlantic*, June 2. https://www.theatlantic.com/entertainment/archive/2016/06/more-likely-to-be-killed-by-a-terrorist-than-to-get-married/485171/.

Glud, E., S.K. Kjaer, R. Troisi, and L.A. Brinton. 1998. Fertility drugs and ovarian cancer. *Epidemilogic Reviews* 20: 237–57. https://www.physio-pedia.com/images/3/36/Fertility_and_ovarian_cancer.pdf.

Lander, Clara. 1971. A dangerous sickness which turned to a spotted fever. *Studies in English Literature, 1500-1900* 11: 89-108.

Lehrer, Jim. Interview with Margaret Edson. 1999. *PBS Newshour*, April 13.

Lo, Bernard. 2012. *Ethical Issues in Clinical Research*: A Practical Guide. New York: Lippincott, Williams, and Wilkins.

Lyall, Sarah. 2001. For Wit, a star who supplies a wit of her own. *New York Times*, March 18. http://www.nytimes.com/2001/03/18/arts/television-radio-for-wit-a-star-who-supplies-a-wit-of-her-own.html.

Marks, Peter. 1998. Science and poetry face death in a hospital room. *New York Times*, September 18.   http://www.nytimes.com/1998/09/18/movies/theater-review-science-and-poetry-face-death-in-a-hospital-room.html.

Mastroianni, Anna C., Ruth Faden, and Daniel Federman. 1994. Women and Health Research: A Report from the Institute of Medicine. Washington: Institute of Medicine.

Modan, Baruch, P. Hartge, G. Hirsh-Yechezkel, A. Chetrit, F. Lubin, U. Beller, G. Ben-Baruch, A. Fishman, J. Menczer, J.P. Struewing, M.A. Tucker, S.M. Ebbers, E. Friedman, B. Piura, and S. Wacholder. 2001. Parity, oral contraceptives, and the risk of ovarian cancer among carriers and noncarriers of a brca1 or brca2 mutation. *New England Journal of Medicine* 345: 235–240. http://www.nejm.org/doi/full/10.1056/NEJM200107263450401#t=article.

Moss, E.L., J. Hollingworth, and T.M. Reynolds. 2005. The role of CA125 in clinical practice. *Journal of Clinical Pathology* 58: 308–312. http://jcp.bmj.com/content/58/3/308.

Museum of London. 2017. Plagues of London. https://www.museumoflondon.org.uk/application/files/5014/5434/6066/london-plagues-1348-1665.pdf. Accessed 22 Feb 2018.

Nicks, Denver. 2015. This was the Hillary Clinton comment that sparked Lena Dunham's awareness. *Time*, Sept 29. http://time.com/4054623/clinton-dunham-tea-cookies/.

Park, Alice. 2014. Lessons from the woman who discovered the BRCA cancer gene. *Time*, June 2. http://time.com/2802156/lessons-from-the-woman-who-discovered-the-brca-cancer-gene/.

Pressley, Nelson. 2000. A teacher's 'wit' and wisdom: Margaret Edson, finding lessons in her sole play. *Washington Post*, February 27.

Rosenthal, M. Sara. 1998. *The Fertility Sourcebook*. New York: McGraw-Hill.

Rosenthal, M. Sara. 2002. *Women Managing Stress*. Toronto: Prentice Hall Canada.

Rosenthal, M. Sara. 2013. The End of Life Experiences of 9/11 Civilians: Death and Dying in The World Trade Center. *Omega Journal of Death and Dying* 67: 323–361.

Rosenthal, M.S., M. Clay, and A. Greer. 2015. The Moral Distress Education Project. www.moraldistressproject.org. Accessed 22 Feb 2018.

Scheibinger, L. 2003. Women's health and clinical trials. *Journal of Clinical Investigation* 112: 973–977. https://www.ncbi.nlm.nih.gov/pmc/articles/PMC198535/.

Television Archive. The September 11th Collection. http://televisionarchive.org/sept11.html. Accessed 19 July 2017.

The Federal Glass Ceiling Commission. 1995. A Solid Investment: Making Full Use of the Nation's Human Capital. Recommendations of the Federal Glass Ceiling Commission. (Report). https://www.dol.gov/dol/aboutdol/history/reich/reports/ceiling2.pdf.

White Junod, Suzanne. 2008. FDA and Clinical Drug Trials: A Short History. In *A Quick Guide to Clinical Trials*, ed. Madhu Davies and Faiz Kerimani, 25–55. Washington: Bioplan, Inc. https://www.fda.gov/AboutFDA/WhatWeDo/History/Overviews/ucm304485.htm.

Wolf, Naomi. 2001. *Missed Conceptions*. New York: Double Day.

Any curriculum that comprises teaching withdrawal of life sustaining treatment or life support must include a discussion of the 1976 Karen Ann Quinlan case (re: Quinlan 1976), which will subsequently lead to the 1990 Nancy Cruzan case (Cruzan V. Director 1990). *Whose Life Is It, Anyway?*—the film adaptation of the 1979 Broadway play—is the film you need to enhance these concepts, while showing students the timeframe in which they were conceived. This film emerges just as the field of "clinical ethics" is born—when debates in the medical literature over mechanical life support and other life sustaining treatments were on "simmer", and clinical ethical principles and frameworks are taking shape in the wake of a significant publication—the first edition of The Principles of BioMedical Ethics (Beauchamp and Childress 1978), in which the Principle of Autonomy and patients' rights to self-determination are now ethically and legally grounded, and the Principles of Beneficence and Non-Maleficence are helping to place limits on overly aggressive care plans. This film is about a car accident victim who was left a quadriplegic and his request to die a natural death because he does not want to be dependent on life sustaining treatment. This film also raises questions about decision-making capacity and informed refusal, but is a classic clinical ethics film that deals more with the concept of withdrawal and withholding of life sustaining treatments and death with dignity. The film challenges different values and perceptions of what it means to have quality of life, and who decides.

The Story Behind *Whose Life Is It, Anyway?*

This film was adapted from a British stage play by Brian Clark. The stage play was originally written as a teleplay, performed on British television on March 12, 1972 (Clark 1974; Gilbert 2006), which was a time frame prior to the U.S. Quinlan case. Clark's 1972 play was likely inspired by a robust debate that was going on in

© Springer International Publishing AG, part of Springer Nature 2018   27
M. S. Rosenthal, *Clinical Ethics on Film*,
https://doi.org/10.1007/978-3-319-90374-3_2

British Parliament from 1969–71 regarding physician-assisted suicide for patients who were suffering from terminal conditions or living under conditions they found objectionable (see further under Social Location). In 1970, Australia passed the first known mandatory seat belt law to specifically reduce the incidence of spinal cord injury (Milne 1985; Schiller and Mobbs 2012), and this may have received considerable attention in the United Kingdom, when Clark was writing the teleplay. The Australia law was likely inspired by a 1965 bestselling book by Ralph Nader, Unsafe at Any Speed (Nader 1965; Jensen 2015), which was a "whistleblowing" non-fiction book that exposed the American automobile industry's willful disregard for consumer safety in its car designs. The book's influence, and Nader's work in this area, is discussed in more detail further on.

Clark then adapted his teleplay into a stage play in 1978 (Speakman 1989), which premiered at the Mermaid Theatre in London, and then opened on Broadway (directed by Michael Lindsay-Hogg) in 1979 to critical acclaim, starring Tom Conti and Jane Asher as the patient, and physician, respectively. The gender roles were then flipped in a 1980 stage revival, in which Mary Tyler Moore played the patient and James Naughton played her doctor. The success of the stage play inspired its adaptation to a feature film, which premiered in December 1981. Since the film's first debut, the stage version was revived for the London stage in 2005 and 2016.

About the Film

The film was adapted for the screen by Brian Clark and Reginald Rose, and directed by John Badham, who made his film directorial debut with *Saturday Night Fever* (1977). The film cast Richard Dreyfuss in the role of patient, Ken Harrison, the sculptor who survived a car crash that caused his spinal cord injury. The *New York Times* film review (Maslin 1981) noted that Dreyfuss' charisma and ability to convey wit and humor with his ongoing banter with the healthcare staff was critical to the film's success, but that in the few scenes without him, the audience missed him. Dreyfuss was at the peak of his career at this phase; he was the youngest actor to win the Academy Award for best actor in 1978 for his role in *The Goodbye Girl* (1977)—a role that depended on his ability to engage in witty repartee. At this point in his career, he had starred in two early Steven Spielberg "blockbuster" hits, *Jaws* (1975) and *Close Encounters of the Third Kind* (1977). But Dreyfuss had succumbed to a cocaine addiction in the late 1970s-early 1980s. He was not well during the making of this film, which has been attributed to his heavy drug use; he could only film for about two or three hours at a time. He reportedly does not have a clear memory of even making the film.

In 1982, a few months after the film's release, Dreyfuss blacked out while driving and his car overturned (Catania and DeSantis 1995; McDonnell 2011). This experience led to his sobriety. He recalled many years later the following: "I was upside down with my head on the pavement with a Mercedes Benz on top of me. And I was held in by a safety belt that I hadn't put on…." In 1995, Dreyfuss was in another crash (not drug-related) when his car ran into a light pole on Ventura Blvd;

he survived yet another car crash in 2016 (TMZ 2016) when he hit a parked car (also not drug-related). Thus, his role as a quadriplegic following a motor vehicle accident was prescient. In his early twenties, Dreyfuss worked as a clerk in a Los Angeles hospital, and so the hospital setting was not unfamiliar to him. When asked why he developed his addiction, he recalled in later years: "[because of the] many of the reasons that young people have been doing stupidly risky things forever…I thought I couldn't be killed, I thought I was immortal, and I thought I would always be smarter and faster…." (McDonnell 2011). The notion of "risk" is an interesting theme to explore when teaching this film, which I discuss further on (see Under Clinical Ethics Issues).

Whose Life is it, Anyway? premiered December 2, 1981 before there were any publicized campaigns against drinking and driving, or any mandated seatbelt laws in the U.S. The film's realistic portrayal of a car crash, and subsequent spinal cord injury resulting in quadriplegia, resonated with audiences—especially when first released around the holidays, in which either alcohol-related, or weather-related car accidents were most common.

Synopsis

This film is about Ken Harrison, a sculptor with a rich, full, artistic life who is in a motor vehicle accident and wakes up a quadriplegic from a severe spinal cord injury. When he begins to understand and appreciate that his condition is life-long, he decides that he does not want to live this way and wants all life sustaining treatments to stop, and to be discharged from the hospital, which will allow his natural death. If this were a patient we had today, we would honor the request. But back in 1981 (and certainly in 1972, when the script debuted on British television), it was not a "standard of care" to withdraw or withhold life support, and most hospitals and doctors would not do it. This becomes the main issue, and hence title of the film: whose decision is it, or *whose life is it, anyway?* The healthcare providers looking after Ken become polarized with this moral dilemma, and Ken takes matters into his own hands and hires an attorney to plead his case to die. In a nutshell: this is about the request for withdrawal/withholding of life support, and the refusal of the hospital to comply with the patient's wishes. However, the character is intensely likable, which creates a moral dilemma for the audience, too. He engages in a "nonstop barrage of one-liners about his sexual incapacity and his vegetable-like-condition. Nurse Rodriguez (Alba Oms) mothers him while Mary Jo (Kaki Hunter), a pretty, young trainee, turns him on. John (Thomas Carter), an orderly with a free spirit, a love of rock music, and a good sense of humor, is the only one who really connects with Ken's concerns." (Brussat and Brussat 2017.) The hospital tries to make an incapacity argument: that Ken is too depressed to make a rational decision. Ultimately, a judge decides Ken's legal fate, and grants his request.

The Social Location of *Whose Life Is It, Anyway?*

When Clark is developing the first teleplay version of the script around 1970–2, the British Parliament was engaging in debate over physician-assisted suicide based on new questions already arising in the medical literature surrounding the ethics of life support (House of Lords 1969; BMA 1971; Voluntary Euthanasia Society 1971; Sommerville 2005). Clark creates a case everyone can relate to: a spinal cord injury following a motor vehicle accident. In other words, a case that involves *risk of daily life* is a case that engages the public's imagination. The discussion of physician-assisted suicide is still premature in the United States at this time, but is raging in Britain, as a Bill is introduced in the British Parliament in 1969, and debated in 1970 and 1971, respectively. This is a freshly post-1960s time frame, and Britain's role in shaping the 1960s cannot be overlooked. Britain had virtually transformed the cultural landscape in the U.S. with the "British Invasion" that included music and fashion. When we think of the "Sixties"—we typically think of the sharp cultural changes that took place beginning around 1966. But those cultural changes began in "Swinging London" (Gilbert 2006). The British influence in music and fashion, such as the mini skirt, bright "psychedelic" colors, models such as Twiggy, were a *cultural export* to the U.S., which then began its own transformation during the late 1960s, including the wide use of recreational drugs, such as LSD. Britain was also ahead with respect to reproductive rights; abortion had been legal in Britain since 1967. Thus, by the early 1970s, when Clark is shaping his first teleplay, he is shaping it within a *British* social context, and not an American one. Prior to the 1960s, London had been considered a relatively bleak and conservative city, associated with the London Blitz, and fighting for its life during World War II. Britain's post-war period was different than that of the U.S. It was rebuilding, as Europe was in a "reconstruction" period. The children of the survivors of the London Blitz were the first generation that had not been drafted, and there was a spirit of liberation that was more pronounced in Britain than in the U.S. at that time, which was actually experiencing a full draft due to the Vietnam War.

But more critically, Britain in its postwar period, started its National Health Service (NHS) in 1948 for the United Kingdom, in which a single-payer, universal healthcare system was established by the British government (the Labour Party at the time) for all its citizens. Thus, questions surrounding resource allocation and patients' rights became part of the public debate earlier. In 1969, a physician-assisted suicide Bill was introduced into the House of Lords. In 1970, the House of Commons debated the issue. Another Bill was introduced into the House of Lords in 1976 on the matter of "passive euthanasia". The following statements were made by debating British parliamentary representatives in 1969, which reveals how much further along Britain was on these issues than the Americans at the time (House of Lords 1969):

One American journalist, who had watched his sister die, wrote that, "spending up to 22 h a day for almost two months, in the hospital room of someone you love, witnessing not the prolongation of life but the long-drawing-out of dying, leaves you looking at such things quite differently to the way you ever had before.

And this:

I do not think I can adequately describe the feelings of the people who have written or spoken to me, and I will not quote them. But it may seem paradoxical that the science of medicine, by the very competence of its resuscitatory techniques, has increased many people's fear of a vegetable existence beyond their normal span. I know that there are objections to this Bill. They are moral and ethical, religious, medical and practical. But I am bound to say to your Lordships that since the Bill was published I have been constantly encouraged and heartened by the sheer number of people, strangers as well as friends, who have wished the Bill well; and the letters I have received have been overwhelmingly—ten to one—in favour of the Bill.

It is a fact that most dying people cling to life. But the purpose of this Bill is to give to others, though they be the minority, the freedom not to live on if they do not or would not choose to. It has not always been considered that one's life is one's own. It used to be thought to be the property of the State, and to some extent it still is the case. Conscription into the Armed Services is one example. But, since the 1961 Suicide Act, to kill oneself is no longer a crime, a felony, and one is free to attempt it without fear of prosecution. So it is not a question of whether people should be given the right to kill themselves—this they already have; it is whether that liberty should be extended, under certain safeguards, to those dying people who are in no position to help themselves and would like to have their lives terminated by the kindly action of medication administered by the agency of another.

The Stage Play: 1978–80

When Clark returns to the script to adapt his teleplay to a stage play in 1978, the Quinlan decision (Re: Quinlan 1976; Sullivan 1976) had occurred, which completely changed the course of medical history in the United States and beyond, giving new meaning to the ethical questions raised in his teleplay, as such questions were now resonating with a post-Quinlan American audience, too. Karen Ann Quinlan, who was adopted (Quinlan 2005) was 21 years old in 1975, and combined alcohol with valium at a party she attended; she had reportedly not eaten for a couple of days prior in order to "diet" so she could fit into her party dress. She collapsed and appeared to have had an anoxic brain injury, which led to a diagnosis of a persistent vegetative state. Her parents wanted the hospital to disconnect her from her ventilator, and the hospital refused, which resulted in the parents suing to have the right to make that decision. The health law context and issues in the Quinlan case are discussed further (see under History of Medicine), but the Quinlan case resonated with parents who were all coping with widespread recreational drug use by their teen or college-age children, a byproduct of the 1960s. For example, the well-known comedian, Carol Burnett, had gone public on the *Mike Douglas Show* with her daughter's drug addiction problem in 1979 to send the message to other parents that they were not alone (Scott 1979). Thus, the Quinlan case was not just about mechanical life support or ventilator support, but about every parent's worse nightmare: dealing with a disastrous outcome based on what was now a "risk of daily life"—watching your college-age children make dumb choices. And that also included drinking and driving, an unspoken theme.

When the Broadway play premieres in 1979, Americans certainly had driving on their minds for other reasons: they were living amidst the 1979 oil crisis (preceded by the 1973 oil crisis under President Nixon) sparked by the political upheaval in the Iran-Iraq region, and the Iranian revolution; Iran had ceased producing and exporting oil during this period, which threw the global oil market into a panic and crisis. Americans were lining up at the gas pump and paying higher prices than they had ever seen. But many were now trading in their gas guzzlers for smaller, economy cars that were more fuel efficient. In response to the 1973 crisis, fuel economy legislation was passed by Congress in 1975 (U.S. Congress 1975) and by 1979, smaller cars (with hatchbacks) were just beginning to be introduced into the market that had better mileage per gallon (e.g. the Ford Pinto; Dodge Dart; and Chevy Nova) but there were also safety concerns with the smaller cars such as the notorious Ford Pinto. There was also a huge shift away from American-made cars; Americans were now purchasing smaller and more fuel-efficient Japanese cars, such as Datsuns (later Nissan), Toyotas and Hondas. Thus, with smaller cars on the road, continuing safety concerns with various cars became top-of-mind when purchasing a car.

In 1980 (during the peak of the Iran Hostage Crisis), when the play was revived with Mary Tyler Moore in the starring role, the playbill was notably sponsored by Beefeater Gin (Playbill 1980), an utterly tone deaf sponsorship for a play about a car crash victim. However, it reflected the disconnection between the rise of motor vehicle accidents with the pervasive advertising and promotion of alcohol. Coinciding with the 1980 version of the play was the establishment of Mothers Against Drunk Drivers (MADD), founded by Candy Lightner in response to the death of her 13-year old daughter who was hit by a drunk driver. MADD was responsible for lobbying to change to the legal drinking age in the U.S. to 21 (MADD 2017).

Car Safety and Seatbelt Laws

In the U.S., a young Ralph Nader was becoming increasingly well known as a critic of the auto industry in the wake of his best-selling book, Unsafe at Any Speed: The Designed in Dangers of the American Automobile (Nader 1965) which excoriates the American car industry for producing unsafe vehicles that magnified, rather than reduced, car crashes and resulting injuries. The book also criticized U.S. automakers for resisting safety features, such as seatbelts. Nader called upon the federal government to regulate the auto industry by enforcing safety features that could reduce preventable death and injury. Ultimately, the book led to the passage of the *National Traffic and Motor Vehicle Safety Act* (1966) as well as to the requirement, in 1968, that all car models be equipped with seatbelts. Yet mandatory seat belt laws were years away. The industry resistance to seatbelts mirrored later industry resistance to mandatory airbags as standard features; airbags had been invented and in use throughout the 1970s as extra features, but the industry resisted it as a standard feature until it was required by law in 1989 (Molotsky 1984).

The play, "Whose Life is it, Anyway?" materializes in the pre-seat belt law era, and certainly pre-airbag era. Although all cars had seatbelts by 1968, *wearing a seatbelt was voluntary* in the United States until 1984, when New York became the first state to require drivers and passengers to wear them. Europe and Australia had passed mandatory seatbelt laws in the 1970s.

Some proposed the argument that mandatory seatbelt laws could lead to riskier or careless driving habits (speeding, etc.), as the seatbelt wearer would feel more emboldened. The argument's proponent was based on 1975 theory about "risk compensation"—in this case—using seat belts—might encourage careless driving (Cohen and Einav 2001).

By the time the Broadway version of the play debuted, so had a new book validating Nader by former General Motors executive, John DeLorean: On a Clear Day You Can See General Motors (DeLorean 1979); DeLorean stated in his book that all of Nader's criticisms of the car industry were valid.

In 1990, *Whose Life is it, Anyway?* had renewed resonance when another landmark Supreme Court decision occurred surrounding a 1983 car crash victim who was not wearing a seatbelt: the Nancy Cruzan case (Cruzan v. Director 1990; Cruzan v. Harmon 1989). Nancy lost control of her 1963 Rambler (a model in which seatbelts were not "standard" until 1968), and was ejected 35 feet through the windshield (Brower and Breu 1989). Her injuries resulted in a persistent vegetative state. In the Cruzan case, the right to withdraw nutrition and hydration was established, which built on the Quinlan decision. The film, in 1981, makes reference to two major issues that surfaced in the Cruzan case: wearing a seatbelt (Ken Harrison asks Dr. Scott if she's planning to "drive home" and tells her to wear a seatbelt) as well as "stopping feeding"—Dr. Emmerson offers stopping of feeding to Ken at the end. (These statements can be plucked out of the film as a natural bridge to a discussion of the Cruzan decision.) It's not so much that the Cruzan case posed unusual questions; it was that the Cruzan case embodied everyday car accident and trauma cases in American hospitals. Had Cruzan been driving a car today, her seatbelt (or later, airbag) would have prevented her injury.

The Right to Die Movement in the U.S.

While Clark's conception of his teleplay was likely in response to the debates in Britain at the time, the societal roots of the American Right-to-Die movement was gaining momentum, which coincided with intense American audience interest in the long-running play and film. Such debates preceded the Quinlan decision, but became more mainstream after Quinlan.

In 1967, the first living will appear in the *Indiana Law Journal*; and a right-to-die bill is introduced by Dr. Walter W. Sackett in Florida's State legislature, which later fails (Death with Dignity 2017). In 1969, a voluntary euthanasia bill is introduced in the Idaho State legislature, which also fails. That same year, Elisabeth

Kubler-Ross publishes <u>On Death and Dying</u> (1969), which was the first substantial qualitative research study on the quality of life of the dying, in which we learned about stages of death: Anger, Denial, Bargaining, Acceptance.

By the 1970s, the patients' rights movement begins to merge with right-to-die initiatives. By 1974, the first hospice opens in New Haven, while a Gallup poll reveals that 53% of Americans are in favor of assisted suicide to lessen suffering. By 1979, in the wake of Quinlan, the timing of the film, *Whose Life is it, Anyway?* finds fertile ground surrounding right-to-die debates in the U.S. Coinciding with the film's release in 1981, the Hemlock Society publishes a do-it-yourself manual for suicide: <u>Let Me Die Before I Wake</u> (Humphry 1981). As the play tours globally at this time, these questions had different cultural debates, ranging from support for euthanasia in Japan, to protests against the play in Germany because it was too close to that country's Third Reich history (Glaap 1989).

The Disability Rights Movement

The play and film, *Whose Life is it, Anyway?* comes out a year after *The Elephant Man* (1980), which is discussed in Chap. 8, and which resonates deeply with the disabled community. I refer readers here to the lengthy section on the historical and social context of disability rights during this timeframe, as similar issues apply here.

Gender Roles in *Whose Life Is It, Anyway?*

For a medical play of its time, it is remarkable that the nascent script in 1972 featured a female physician (Dr. Scott) at all, which was a consistent character throughout all versions of the play and later screenplay (Clark 1974). (The role of women in medicine, and medical school training with respect to films about Professionalism and Humanism is discussed in <u>Healthcare Ethics on Film</u>.) It is also of note that the gender roles of the patient and Dr. Scott switch back and forth in some productions of the play (Bench Theatre 1983; The Broadway League 1980). In some reviews of the 1980 revival of the play with Mary Tyler Moore (who also starred in *Ordinary People* that year, playing a tortured mother of a teen suicide victim), there was speculation that Moore's star power was an economic choice (Nakano 2017). I suggest that switching the patient's gender was a direct response to the Quinlan decision, in which a woman's body becomes the center of a polarizing debate over end of life decisions and patient autonomy. In 1980, in the early wake of feminist bioethics, a female patient fighting to make decisions about her own body was not an accident, but a reflection of where audiences and debates were at that time. Male reviewers were not that pleased with the gender role reversal; a February 25 1980 review in the *New York Post* quipped: "Samson could play Deliliah, and Delilah could play Samson, but who would have the scissors.

Unisex plays can only work with unisex playwrights and unisex audiences" (Glaap 1989). Indeed, in the revival of the stage play in 2005 and 2016, gender role reversals between the patient and Dr. Scott continue, as the debates and controversies over women's bodies have continued to rage. Brian Clark stated: "a woman taking over is a modest contribution to the feminist cause…in which I'm a believer" (Glaap 1989). Clark also wanted the patient's plight to have a universal appeal.

History of Medicine Context: 1970–81

In exploring the history of medicine context for the teleplay, stage play and film, the issues revolve around growing concerns over resource allocation with respect to futility and end of life care in intensive care units (ICUs), which coincide with expanded definitions of death when the Harvard Brain Death Criteria is published in 1968 (JAMA 1968). By defining parameters for death by neurological criteria, some problems were resolved, while other problems were created, as bedside practices and communication with families surrounding brain death continue to be problematic.

Additionally, landmark U.S. health law decisions begin to establish standards for informed consent, refusal of treatment, including life-sustaining treatment, as well as the establishment of Hospital Ethics Committees. By the time the stage play debuts on Broadway, the first core principles of biomedical ethics are published in two works: *The Belmont Report* (National Commission 1979), and in the first textbook on the topic, which continues to be the dominant text in the field. Central to the tenets of the Principle of Autonomy are the concepts of competency and decision-making capacity, which are major themes in the film. Finally, another area to discuss in this context is the history of spinal cord injuries and rehabilitation medicine as a field; the issue of clinical depression as a complication of spinal cord injury has never been adequately resolved, but raises interesting questions as to whether the character in the film has the capacity to decide.

Luce and White (2009) note the following in the journal *Critical Care Clinics*, which nicely introduces the ICU history of medicine context:

> Few outcome studies were actually available in the 1950s, 1960s, and 1970s, other than those showing improved survival due to the detection and correction of arrhythmias after myocardial infarction. Nevertheless, because they housed impressive innovations that could reverse physiological dysfunction and sustain life, ICUs became a standard of care. In the process, physicians and others who practiced in the units witnessed increasing conflicts among the ethical principles applied there.

Thus, questions about sustaining life when there was no quality of life led to the common term "medical futility" only recently revised to less normative vernacular of "medically inappropriate" or "non-beneficial" treatments.

The Harvard Brain Death Criteria

Whose Life is it, Anyway? emerges at a time where death and life were redefined by the concept of neurological death, or Brain Death. In 1968, an ad hoc committee at Harvard Medical School "reexamined the definition of brain death and defined irreversible coma, or brain death, as unresponsiveness and lack of receptivity, the absence of movement and breathing, the absence of brain-stem reflexes, and coma whose cause has been identified." (Goila and Pawar 2009). Prior to these criteria, the term was used loosely, which often encompassed a range of neurological injuries, including PVS. In the film, Ken Harrison argues that he does not meet the classic definitions of biological life; we are forced to agree that the only part that was left intact from the accident is his brain. When the play emerges, it is during the pioneering phase of "brain death" criteria, which often became a chaotic bedside blunder, as many healthcare providers and families found it difficult to grasp that patients who are "brain dead" are just as dead as patients who are dead by cardiopulmonary criteria. We began to learn that additional guidelines were needed about how to confirm and explain death that is determined by neurological criteria. Often, families would be asked for permission to remove "life support" from a "brain dead" patient, which was very confusing, and families would typically refuse, or threaten to sue. On the flip side, many healthcare providers held firm convictions that removal of any life sustaining treatment, even with a diagnosis of brain death, was in violation of their personal values—particularly if they were Catholic. There were also conflicting opinions as to whether certain patients were wrongly pronounced brain dead, while many had religious and moral objections to this definition. The American Academy of Neurology introduced practice guidelines in 1981 (JAMA 1981) for declaring death by neurological criteria, which can be done based on clinical exam alone. Confirmatory tests, such as EEG or blood flow studies are not necessary unless the clinical exam is inconclusive for some reason.

When teaching this film, revisiting the definitions of life and death force reflection over qualitative versus quantitative definitions of life. Brain death is thus recognition that there is no personhood left. In the case of quadriplegia, even though the body cannot live without life-sustaining treatments, quality of life is literally a "state of mind", and for these reasons, Ken Harrison's decision is disturbing because in his case, one can argue that a vibrant mind without a mobile body is still a life worth living. He thinks not, given that use of his hands defined his sense of self and personhood.

1972: Year of Informed Consent

The legal doctrine of informed consent began to take shape as early as 1914 with a case that involved a total hysterectomy without consent in *Schloendorff v. Society of New York Hospital* (1914), which introduced the concept of patient "self determination". Other precedent-setting cases include: *Salgo v. Leland Stanford Jr. University Board of Trustees* (1957), which introduced the term "informed consent" and criteria for disclosure; and *Natanson v. Kline* (1960), which first proposed that failure to disclose risks constituted negligence. But in 1972, when the teleplay was first performed, it coincided with three landmark U.S. informed consent cases: *Canterbury v. Spence* (1972), *Cobbs v. Grant* (1972), and *Wilkinson v. Vesey* (1972) most firmly established informed consent as a legal requirement in healthcare. All three cases involved patients who were not sufficiently informed about the risks of procedures. In *Canterbury v. Spence* (1972) the court stated that "informed consent is a basic social policy...." In *Cobbs v. Grant* (1972), the court emphasized "a duty of reasonable disclosure of the available choices ...[and] the dangers inherently and potentially involved in each." *In Wilkinson v. Vesey* (1972), the court stated: "a physician is bound to disclose all the known material risks peculiar to the proposed procedure."

1976: The Quinlan Decision and Withdrawal of Life Support

As discussed earlier, the Quinlan case (re: Quinlan 1976) was about a 21 year-old woman who collapsed at a party after combining alcohol with valium; she had an anoxic brain injury, and remained in a persistent vegetative state. When it was clear that there was no hope of improvement, the parents requested the ventilator be withdrawn, and the hospital refused because at the time, it was believed to be a criminal act. The courts ruled in favor of the parents, and allowed withdrawal of life sustaining treatment in this circumstance. The case changed medicine forever (Angell 1995).

Notwithstanding the pre-Quinlan influences that informed the teleplay and stage play version of "Whose Life is it, Anyway?", by the time the film was being contemplated, the Quinlan decision was big news (Sullivan 1976), and it was clear that the film would resonate with audiences who had read about the case. By 1977, both a made-for-television NBC film and a book about the case were released, which told the parents' story (Kron 1977).

In the "Decision Scene" of *Whose Life is it, Anyway?* the judge even calls his assistant and asks her to look up the Karen Ann Quinlan case. And so, as the chapter title states, the *film, Whose Life is it, Anyway?*, is really about the Quinlan case: withdrawal of life sustaining treatment. But the issues are also about the Cruzan case, even though it was a decade after the film's release.

I suggest that this film should be a companion to teaching about Quinlan because the film clearly demonstrates what hospital culture was like when the case took place. In the made for TV film on the Quinlan case, the last scene ends dramatically as the parents watch the ventilator being withdrawn. In *Whose Life is it, Anyway?* the film ends with the curtain being drawn around the patient for privacy; the character, though not on a ventilator, requires dialysis and other hospital treatments to live.

The details of the Quinlan case involve two issues: whether hospitals can withdraw life support without being charged with a crime; and whether the patient (or surrogate) has the right to make such a request. The New Jersey Supreme Court held that patients or surrogates have the right to refuse life-sustaining treatment in cases where there is no hope of recovery. Quinlan continued to breathe on her own until 1985 (Washington Post 1985; Hornblower 1985; McFadden 1985), nine years after she was withdrawn from her ventilator. Her life continued because nutrition and hydration continued. This was settled with the Nancy Cruzan case (see further).

Thus, in the Quinlan decision, the court ruled in the surrogates' favor to remove the ventilator, citing the irreversible nature of her vegetative state. By extension, withdrawing life support (or life-sustaining treatment) in other similar cases became a standard—even for patients who could make the request themselves because they felt they had no quality of life.

This is also the first case in which the courts suggest a hospital "ethics committee", which after the Cruzan case (see further), became a requirement for most hospital accreditation bodies in the United States. The Decision states (re Quinlan 1976):

> ...[The attending physicians] shall consult with the **hospital 'Ethics Committee'** or like body of the institution in which Karen is then hospitalized. If that consultative body agrees that there is no reasonable possibility of Karen's ever emerging from her present comatose condition to a cognitive, sapient state, the present life-support system may be withdrawn and said action shall be without any civil or criminal liability therefore, on the past of any participant, whether guardian, physician, hospital or others.

1978–1979: The Year of Core Clinical Ethics Principles

In the post-Quinlan era, when the London and American stage plays emerge, the timing coincides with the formation of the National Commission for Human Subjects Protections' publication of *The Belmont Report* (1979), which are agreed-upon principles for medical research involving human subjects (Beauchamp 2010). The formation of the National Commission and this report is discussed more in Chaps. 9 and 10. But the same author of the Belmont Report principles was simultaneously working on a much deeper treatise on core ethical principles for medicine: The Principles of Biomedical Ethics (Beauchamp and Childress 1978). This work took hold as the dominant framework for everyday practitioners looking for an applied set of ethical principles for practice (known as the Four Principles). Although there are continuing critiques by other philosophers about whether the Principlism approach sufficiently accounts for all ethical behaviors and issues in medicine, the four principles outlined by Beauchamp and Childress are well-defended and practical foundations for clinical ethics guidelines and frameworks. *Whose Life is it, Anyway?* can be said to be a clinical ethics film in "real time"—it raised questions about patient autonomy and patient rights just as clinical ethics was emerging as its own field, with its own codified frameworks that supported the legal doctrine of informed consent, and the patient's rights to refuse treatments.

The Cruzan Case

Although this film is pre-Cruzan, the reason *Whose Life is it, Anyway?* remains fresh and relevant is because of the Nancy Cruzan case (Cruzan v. Director 1990). This 1990 Missouri decision established the right of competent patients to refuse life-sustaining treatments that included nutrition and hydration (Schwartz 1990). In this case, a 30 year-old car accident victim suffered a head trauma that left her in a persistent vegetative state, although she was able to breathe on her own. Her parents agreed to a feeding tube (gastrostomy tube) in the belief that she had a chance to recover. However, when it became clear there was no chance of improvement, they requested withdrawal of the feeding tube, and the hospital refused. Eventually, the courts ruled in the parents' favor. The case also established that such refusals could be honored by a substituted judgment made by a surrogate, based on *known or previously stated preferences* of the patient. In Cruzan, the courts recognized that the parents provided "clear and convincing evidence" about what their daughter would have wanted if she were competent because she had prior conversations about not wanting to live in a compromised state (she indicated she wanted a "half normal life"); it is likely these wishes were triggered by a discussion of the Quinlan case. The Cruzan case also established that nutrition and hydration count as medical treatment, and can be withheld. The case also paved the way for the *Patient Self-Determination Act* (1990), which requires all hospitals to discuss Advance Directives with patients.

The State of Spinal Cord Injury and Rehab Medicine

If this film is being used in a medical school context, a discussion about the history of spinal cord injury would be prudent. Prior to the mid-twentieth century, most people with this injury died within a few days. By the 1940s and 1950s, life expectancy was greatly extended with better techniques to drain fluids, prevent urinary tract and other infections, as well as better surgical innovations to deal with certain complications. Spinal cord injury was considered to be more of an orphan condition that received less research attention than other diseases until more recently (Donovan 2007). The main cause of this injury is trauma from a car accident or, among veterans, a bullet wound (in this era, spinal cord injury was a signature war wound of the Vietnam War). The American Spinal Injury Association had formed in 1973. That same year, the introduction of the computed tomography (CT) scan was an important step in properly evaluating spinal cord injuries before magnetic resonance imaging (MRI) was invented. There was recognition in the trauma field that seat belts were considered a major form of prevention, which eventually led to mandatory seatbelt laws. (Similar debates over helmet laws and brain injury abound.) There was also recognition that depression was a major consequence of spinal cord injuries (especially in quadriplegia, coined in 1881, and previously called "cervical paraplegia"). The main reforms in spinal cord injuries were in tool-making, better medical treatments for depression, as well as psychosocial therapies to improve sense of wellbeing and quality of life. But some of the important tools were legislative. For example, although this film takes place prior to the passage of the *Americans with Disabilities Act* (1990), the *Rehabilitation Act* was passed in 1973, which prohibits discrimination against disabled persons in the workplace. Schiller and Mobbs (2012) note: "The focus of spinal cord injury advocacy would slowly shift from the medical aim of achieving survival to the political aim of achieving social equality."

In the film, an important scene that reveals the recognition of treating the psychosocial wounds of spinal cord injury is in the encounter between the Social Worker and Ken Harrison; in this scene, she wants to address his depression, and introduce him to all the new tools available, such as the reading machines, to which he asks if there is a book entitled: "Sculpting with No Hands". The joke exposes the Social Worker's inability to "go with the flow" and relate to her angry patients, and she becomes overly condescending to cover for her own sense of helplessness to really "fix" the depression—which could only be fixed by restoring the spinal cord.

Ultimately, spinal cord injury care requires caring for the psychosocial needs of patients, but truth-telling (discussed under Clinical Ethics issues) is the honest broker that helps patients to decide whether they want to live within the limitations of this condition. One could use the film to juxtapose past limitations with current improvements that enable better social interaction and mobility when considering a curriculum for teaching this film.

Clinical Ethics Issues

The lion's share of clinical ethics issues in this film focus on patient autonomy and the right to refuse treatment. The essential scene that speaks volumes is a confrontation between the patient and Dr. Emmerson, who forces a mood-altering drug on the patient despite his clear objection. "Do *not* put that fucking thing in my arm!" he says, to which the doctor replies that the patient is "depressed" and "can't decide" and proceeds with the injection against the explicit refusal. The doctor explains that the unwanted medication will help the patient sleep, to which the patient retorts: "I don't want to goddamn sleep; I want to goddamn *think*!." The focus of the entire conflict lies in this scene, in which there is an attempt to paint the patient's refusal as being irrational due to depression. The scene also depicts the paternalistic attitudes that reigned in 1981. To show how far we've come, this scene can be used as a snapshot of the uphill fight for patient autonomy.

At the same time, truth-telling is also highlighted in this film. The legal doctrine of informed consent, housed under the Principle of Autonomy, requires "truth-telling"—a *full disclosure* of all treatment options, their associated risks and benefits. The frank disclosure to the patient that he will never improve beyond his current status is what leads the patient to make an informed decision: a decision to ultimately request that all life-sustaining treatments be withdrawn. The patient makes clear: "I am not asking anyone to kill me; I am merely asking to be discharged."

Thus, focusing on the ethical dilemmas of what "informed refusal" (versus consent) actually means, and what withholding/withdrawal of treatment actually means, should inform the clinical ethics curriculum taught with this film, which I focus on in this section.

Refusal of Treatment and Decision-Making Capacity

Every clinical ethicist has received an ethics consultation surrounding a patient's refusal of treatment, and a subsequent questioning of capacity. As the clinical ethicist begins to "work up" the case, s/he is paged again with the requestor insisting: "It's okay; we can cancel the consult; the patient has now agreed to treatment." Unfortunately, there is a common belief that refusing treatment is typically irrational and a sign of incapacity; yet, when these same patients "agree", no one seems to be concerned about capacity anymore. *Whose Life is it, Anyway?* forces an unpacking of this common hospital scenario.

Informed refusal, like informed consent, demands that patients have the *capacity and competency to consent or refuse,* meaning that patients must be considered to meet not just the legal standard of competence (autonomous adults or autonomous emancipated minors) but have decision-making capacity. Finally, for any refusal or consent to be valid, it needs to be *voluntary.*

Decision-making capacity is a medical assessment any Attending physician can make based on the U-ARE criteria: Understanding, Appreciation, Rationality and Expression of a choice. This criteria was established in the late 1970s (Roth et al. 1977), and was known when this film was made. Part 2 of this book is dedicated to the concepts of capacity and competency, and so I refer readers to the more detailed sections on decision-making capacity in that portion of the book. But for the purposes of this film, the fact that the patient might be "depressed", despite clearly meeting the U-ARE criteria, is the crux of the hospital's argument. In the film, two psychiatric opinions are sought which disagree over the role of depression with respect to capacity. In actual cases of spinal cord injury, depression is a problem, and may interfere with rationality as patients require some time to adjust to their new status. At the same time, if the condition itself is the main reason for depression, which leads to chronic existential suffering, this presents the case for the "rational suicide". Depression is not always a barrier to decision-making capacity yet, sometimes it is, as there are adjustment periods in trauma patients.

In the film, the patient states in his legal hearing: "Well, I am completely paralyzed. So if I weren't depressed, I'd be crazy." He goes on to discuss a case of existential suffering that makes clear that continued life support will not meet the criteria for beneficence: maximizing benefit and minimizing harm.

From a clinical ethics framework, a more nuanced position, which may not always be feasible, would be a time-negotiated management plan, in which the patient is assured that his/her requests for withdrawal of treatment will be honored, and are valid. However, the patient should be offered "time" as an option first, and assured that the decision does not have to be made immediately, or is, in any way, an "emergency". Consistency in refusal of treatment is an important marker in evaluating whether the patient's mood is interfering with decision-making. In patients who can walk, refusal of treatment is typically resolved when the patient leaves the hospital Against Medical Advice (AMA), and such is documented in the medical record. AMA discharges do not always involve capacitated patients, but, as discussed more in Chap. 6, on psychiatry ethics (with the film, *One Flew Over the Cuckoo's Nest*), there are also limitations on involuntary hospitalizations. Spinal cord injury patients such as Ken Harrison cannot walk out of the hospital; although he has not been committed to "involuntary" hospitalization as a psychiatry patient, he is nevertheless being held as a "ward" against his will. Thus, when his lawyer, Carter Hill, proposes a "writ of habeas corpus" (a process demanding evidence of cause for holding a person in custody against his/her will), the parallel of hospitalization and imprisonment is raised, and is the main driver of his withdrawal/withholding request.

Withdrawal of Life Support/Withholding of Life-Sustaining Treatments

In the film, Ken Harrison emphasizes that he is not asking to be "killed" nor is he asking anyone to assist in his suicide. He points out that he is merely "asking to be discharged" to which the judge replies: "Which will kill you!" Ken Harrison next states: "Which is exactly my point." This nicely summarizes the moral distinction between physician-assisted suicide and withdrawal or withholding of life-sustaining treatments. In the latter, we are merely permitting a natural death by non-interference in the natural progression of a terminal condition. In the former, we are performing some sort of action that aids in dying. For these reasons, since Quinlan and, especially, Cruzan, there has been widespread acceptance that withdrawal/withholding of life-sustaining treatment supports the Principles of Beneficence and Non-Maleficence. In the late twentieth century, the common ethical dilemma typically involved patients or their families making these requests and finding resistance from healthcare providers to honor the request. In one landmark case (Barber v. Superior Court 1983), two physicians were actually charged with murder when they withdrew life support on a patient who was comatose at the request of the family. The decision stated that: "A physician's omission to continue treatment where such treatment has proven ineffective, regardless of the physician's knowledge that the patient would die, is not a failure to perform a legal duty and therefore the physician cannot be held liable for murder." However, post-Barber, practitioners no longer feared legal ramifications to actively suggesting withdrawal/withholding of life support to families.

In the twenty-first century, we find ourselves with the reverse dilemma: practitioners wanting to withdraw/withhold life-sustaining treatment in "futility" cases surrounding incapacitated patients, and their surrogates refusing to agree, insisting on continuation of all aggressive life support. Some countries and a few U.S. states even have policies in which unilateral withdrawal/withholding of treatment in cases where such treatments merely prolong death, are performed against the family's wishes, which has prompted legal action. The courts have often ruled in these cases in favor of the families, except in cases of brain death. Conflict resolution has become the management plan when there is disagreement over the "goals of care"—allowing a natural death in an irreversible, untreatable condition in which there is no quality of life, or prolonging somatic support without any evidence of personhood. Such conflicts typically revolve around subjective interpretations and worldviews of what "quality of life" or "futility" means. Brody and Halevy (1995) frame different types of futility as physiological, imminent demise, lethal condition, or qualitative. Quantitative futility is where the likelihood that an intervention will benefit the patient is exceedingly poor, while qualitative futility is where the quality of benefit an intervention will produce is exceedingly poor (Schneiderman et al. 1990) Thus, the determination of "futility" may be subjective or objective, depending on the perspective. In this film, the patient argues that he has qualitative futility, but the fact that the patient demonstrates superior capacity, and a sharp mind convinces his healthcare providers that their interventions are beneficial.

Paternalism and Professionalism

This film provides a fertile launch pad for discussing professionalism, paternalism (Dr. Emerson smokes on-the-job) and boundary crossing (between Dr. Scott and the patient). There are also scenes in which the nurses appear to "go along" with continuous sexual innuendo, while a subplot between a nurse and orderly is not politically correct today, but nonetheless probable in its time. Finally, the "pot-smoking" scene, in which the dating couple of an orderly and nurse take the patient out for a "night on the town" is not at all realistic. With few exceptions, the behaviors were accurate in its day, and it's worth discussing the "House of God" era of medicine when there were no workplace laws against smoking, sexual harassment, and few professional guidelines surrounding patient boundaries. (For a much longer discussion on the issues of professionalism and humanism, see Healthcare Ethics on Film.)

"Risk of Daily Life"

Undoubtedly, most educators will be teaching this film to students or trainees in their twenties, who are mandated by law to wear a seatbelt, and whose cars are equipped with airbags—far safer than the car Nancy Cruzan was driving. But we still see many motor vehicle accident-related trauma with distracted driving (texting), Driving Under the Influence (DUI), and even optional helmet use on motorcycles. The film, at its base, is about the consequences of a car accident, which is a "risk of daily life", and reams of statistics demonstrate that people in their twenties take more risks. Absent a car crash, spinal cord injury is generally the result of some sort of accident—falling off horses (the case of Christopher Reeves) and bar stools (a common alcohol-related trauma) are also common causes. While "cutting room floor" decisions will need to be made about which issue to focus on

in this film, the discussion may veer into the meaning of risk. In the wake of
Nader's work on car safety (see above), several philosophers used cars and car
accidents as laboratories in which to analyze what societies would tolerate in terms
of risk. One of my mentors, a philosophy professor in the 1970s, used to give his
class the following question to debate, which has its origins in philosophy of risk
arguments (Nozick 1974; Hansson 2014):

> An inventor proposes a "machine of great convenience" that will change our lives in very
> positive, and stunning ways. The catch is, every year, about a million people will need to be
> sacrificed for this convenience, and about 50 million people will be disabled in order for us
> to use this machine. Are these numbers acceptable to you, in order to have a convenience?

Then a debate ensues in class in which some students will clearly state that these
levels of sacrifice are completely unacceptable. After about 15–20 min, you reveal
the machine is an automobile. And thus begins the bioethics discussion about what
"risk of daily life" truly means, as life is certainly not risk-free, and most trauma is
due to accidents of daily life, or the result of choices to "risk even more". This all
circles back again to Quinlan and Cruzan. A 21-year old "risks more" when she
mixes alcohol and drugs. A 30-year old drives a 20-year old car in 1983 on a dark
rainy night that didn't have any seatbelts.

Conclusions

As I've stressed throughout this chapter, *Whose Life is it, Anyway?* should be
considered a pedagogical enhancement to the teaching of Quinlan, Cruzan and
withdrawal/withholding of life-sustaining treatments; it belongs in any section of a
bioethics course that wants to focus on patient autonomy, end of life, or death and
dying; it is also an appropriate film in a health law class where "Re: Quinlan" is on
the syllabus. This film is also an important enhancement to discussion of death with
dignity laws, and the distinctions between physician aid in dying, and merely
permitting a natural death by removing impediments to that process. Finally, I
conclude this chapter with a reminder of the last line of the film, in which Dr.
Emerson offers to have the patient stay in his room in privacy until he dies—which
Ken agrees would clearly make things easier for him. When Ken asks Emerson why
he is being so accommodating, considering his opposition to the plan, he answers:
"In case you change your mind". And sometimes, patients do.

Film Stats and Trivia

- The original theatrical poster for this film had the tagline: "The Ultimate Drama
  in the human comedy", which was a very deceptive descriptor for a film about
  the request to die.

- Richard Dreyfuss made his film debut was in the film, *The Graduate* (1967), in which he has one line: "Do you want me to get the cops? (When Elaine screams in Benjamin's rented room.)
- Director, John Badham wanted to shoot the film in black and white, but MGM did not like the idea; it proposed that Badham shoot in color and produce one negative in black and white, which it could test. The black and white version had tested better, but the head of MGM preferred to release the color version instead.
- The character of Ken's dancer girlfriend was played by the real dancer, Janet Eilber, who was with the Martha Graham dance troupe.

From the Theatrical Poster

Director: John Badham
Producer: Lawrence P. Bachman
Writer: Brian Clark, Play; Reginald Rose, Screenplay
Starring: Richard Dreyfuss, John Cassavetes, Christine Lahti, Bob Balaban
Music: Arthur B. Rubinstein
Cinematography: Mario Tosi
Editor: Frank Morriss
Production: Metro-Goldwyn-Mayer
Distributor: United Artists
Release Date: December 2, 1981

References

Ad-Hoc Committee of Harvard Medical School. 1968. A definition of irreversible coma: report of the Ad-Hoc Committee of Harvard Medical School to examine the definition of brain death. *Journal of the American Medical Association* 205: 337–40.

Angell, M. 1995. After Quinlan: The dilemma of the persistent vegetative state. *New England Journal of Medicine* 330: 1524–25. http://www.nejm.org/doi/full/10.1056/nejm199405263302110.

Barbar v. Superior Court of California, 195 CAL. R. 484 CT. APP. (1983).

Beauchamp, Tom L. 2010. The origins and evolution of The Belmont Report. In *Standing on Principles* Tom L. Beauchamp, 3–18. New York: Oxford University Press.

Beauchamp, Tom L., and James L. Childress. 1978. *The Principles of Biomedical Ethics*, First Edition. New York: Oxford University Press.

Bench Theatre. 1983. Whose Life is it, Anyway? playbill. https://www.benchtheatre.org.uk/plays80s/whoselife.php. Accessed 22 Feb 2018.

British Medical Association. 1971. The Problem of Euthanasia. (Report) (Note: these reports are referenced here: http://journals.sagepub.com/doi/pdf/10.1177/002581727304100103.). London: British Medical Association.

Brody, Baruch, and Amir Halevy. 1995. Is futility a futile concept? Journal of Medicine and Philosophy 20:123–144.

Brower, Montgomery, and Giovanna Breu. 1989. Nancy Cruzan's parents want to let her die—and are taking the case to the supreme court. *People*, December 4. http://people.com/archive/nancy-cruzans-parents-want-to-let-her-die-and-are-taking-the-case-to-the-supreme-court-vol-32-no-23/.

Brussat, F., and M.A. Brussat. 2017. Film review: Whose Life Is It Anyway? *Spirituality and Practice*. http://www.spiritualityandpractice.com/films/reviews/view/3864. Accessed 19 July 2017.

Canterbury v. Spence, 464 F.2d 772 (D.C. Cir. 1972).

Catania, Sara and Jeannette DeSantis. 1995. Richard Drefuss injured as car hits pole in studio. *Los Angeles Times*, January 16. http://articles.latimes.com/1995-01-16/local/me-20684_1_richard-dreyfuss.

Clark, Brian. 1974. *Whose Life Is It Anyway? A Full-Length Play*, 1st ed. Chicago: Dramatic Publishing Co.

Cobbs v. Grant, 8 Cal. 3d 229, 502 P.2d 1, 104 Cal. Rptr. 505 (Cal. 1972).

Cohen, A., and L. Einav. 2001. The effects of mandatory seatbelt laws on driving behavior and traffic fatalities. Center Discussion Paper No. 341. Harvard Law Online. http://www.law.harvard.edu/programs/olin_center/papers/pdf/341.pdf. Accessed 24 Feb 2018.

Cruzan v. Director, Missouri Department of Health, 497, U.S. 261 (1990).

Cruzan v. Harmon, 760 S.W. 2d 408 (Mo. 1988), cert. granted, 109 S. Ct 3240 (1989).

Death With Dignity. 2017. Chronology of assisted dying. https://www.deathwithdignity.org/assisted-dying-chronology/. Accessed 19 Jul 2017.

DeLorean, John. 1979. *On a Clear Day You Can See General Motors*. New York: Wright Enterprises.

Donovan, W.H. 2007. Spinal cord injury—past, present, and future. *J Spinal Cord Med*. 30: 85–100. https://www.ncbi.nlm.nih.gov/pmc/articles/PMC2031949/.

Gilbert, D. 2006. The Youngest Legend in History: Cultures of Consumption and the Mythologies of Swinging London. *The London Journal* 31:1–14. http://dx.doi.org/10.1179/174963206X113089.

Glaap, Albert-Reiner. 1989. Whose Life is it Anyway? In London and on Broadway: a contrastive analysis of the British and American versions of Brian Clark's Play. In *The Play Out of Context: Transferring Plays from Culture to Culture*, eds. Hanna Scolnicov and Peter Holland, 214–26. London: Cambridge University Press.

Goila, A.K., and M. Pawar. 2009. The Diagnosis of brain death. *Indian Journal of Critical Care Medicine* 13: 7–11. https://www.ncbi.nlm.nih.gov/pmc/articles/PMC2772257/.

Guidelines for the Determination of Death. 1981. Report of the medical consultants on the diagnosis of death to the President's commission for the study of ethical problems in medicine and biochemical and behavioral research. *Journal of the American Medical Association* 246: 2184–2186.

Hansson, Sven Ove. 2014. Risk. In *The Stanford Encyclopedia of Philosophy* eds. Edward N. Zalta. https://plato.stanford.edu/entries/risk.

Hornblower, Margaret. 1985. Quinlan left legacy to era of technology. *Washington Post* June 13. https://www.washingtonpost.com/archive/politics/1985/06/13/quinlan-left-legacy-to-era-of-technology/773e0ae5-dcb6-4d12-add0-31769dcc6888/?utm_term=.19e890157460.

House of Lords. 1969. Voluntary Euthanasia Bill [H.L.] HL Deb 25 March vol 300. cc1143–254. http://hansard.millbanksystems.com/lords/1969/mar/25/voluntary-euthanasia-bill-hl.

Humphry, Derek. 1981. *Let Me Die Before I Wake*. Santa Monica: Hemlock Society.

In re Quinlan 70 N.J. 10,355 A2d 647, cert. denied sub nom. Garger v New Jersey, 429 U.S. 922 (1976).

Jensen, Christopher. 2015. 50 Years ago, 'Unsafe at Any Speed' shook the auto world. *New York Times,* November 26. https://www.nytimes.com/2015/11/27/automobiles/50-years-ago-unsafe-at-any-speed-shook-the-auto-world.html.

Karen Ann Quinlan dies at age 31: coma case prompted historic ruling. 1985. *Washington Post,* June 12.

Kron, Joan. 1977. As Karen Ann Quinlan lives on in a coma, a new book and tv film tell her story. *New York Times*, September 24. http://www.nytimes.com/1977/09/24/archives/as-karen-ann-quinlan-lives-on-in-a-coma-a-new-book and-tv-film-tell.html?_r=0.

Kubler-Ross, Elizabeth. 1969. *On Death and Dying.* The Macmillan Company: New York.

Luce, J.M., and D.B. White. 2009. A history of ethics and law in the intensive care unit. *Critical Care Clinics* 25:221–237. https://www.ncbi.nlm.nih.gov/pmc/articles/PMC2679963/.

Maslin, Janet. 1981. Movie Review: Whose Life Is It Anyway? *New York Times*, December 2. http://www.nytimes.com/movie/review?res=9906E0DA113BF931A35751C1A967948260.

McDonnell, Brandy. 2011. Richard Dreyfuss recalls his journey to sobriety at an Oklahoma City event. *The Oklahoman*, October 12.

McFadden, Robert D. 1985. Karen Ann Quinlan, 31, dies; focus of '76 right to die case. *New York Times*, June 12. http://www.nytimes.com/1985/06/12/nyregion/karen-ann-quinlan-31-dies-focus-of-76-right-to-die-case.html?pagewanted=all.

Milne, P.W. 1985. Fitting And Wearing Of Seat Belts In Australia: The History Of A Successful Countermeasure. Report for the Department of Transport. Canberra: Australian Government Publishing Service. https://infrastructure.gov.au/roads/safety/publications/1985/pdf/Belt_Analysis_4.pdf.

Molotsky, Irvin. 1984. U.S. sets '89 date for car air bags but gives choice. *New York Times*, July 12. http://www.nytimes.com/1984/07/12/us/us-sets-89-date-for-car-air-bags-but-gives-choice.html.

Mothers Against Drunk Driving. 2017. History. MADD.org. https://www.madd.org/history/. Accessed 22 Feb 2018.

Natanson v. Klein, 350 P.2d 1093 (1960).

Nader, Ralph. 1965. *Unsafe at Any Speed: The Designed-In Dangers of the American Automobile.* New York: Grossman Publishers.

Nakano, Craig. 2017. Why Mary Tyler Moore received her special Tony award in 1980. *Los Angeles Times*, January 26. http://www.latimes.com/entertainment/arts/la-et-cm-mary-tyler-moore-tony-20170126-htmlstory.html.

Nozick, Robert. 1974. *Anarchy, State, and Utopia*, New York: Basic Books.

Playbill, 1980. Whose Life Is It Anyway?, theatrical playbill. Posted to: http://www.playbill.com/playbillpagegallery/inside-playbill?asset=00000150-aea6-d936-a7fd-eef62efd0004&type=InsidePlaybill&slide=1 and http://www.playbill.com/production/whose-life-is-it-anyway-royale-theatre-vault-0000010212.

Quinlan, Julia D. 2005. *My Joy, My Sorrow: Karen Ann's Mother Remembers.* Cincinnati: St. Anthony Messenger Press.

Richard Dreyfuss: close encounter with produce truck. 2016. *TMZ*, September 4. Posted to: http://www.tmz.com/2016/09/04/richard-dreyfuss-car-accident-truck/.

Roth, Loren H., Alan Meisel, and Charles W. Lidz. 1977. Tests of competency to consent to treatment. American Journal of Psychiatry 134:279–284.

Salgo v. Leland Stanford Jr. University Board of Trustees, 317P.2d 170,181 (1957).

Schiller, M.D., and R.J. Mobbs. 2012. The historical evolution of the management of spinal cord injury. *J Clin Neurosci* 19:1348–53. https://doi.org/10.1016/j.jocn.2012.03.002.

Schloendorff v. Society of New York Hospital 211 N.Y. 125, 105 N.E. 92 (1914).

Schneiderman Lawrence J., Nancy S. Jecker, and Albert R. Jonsen. 1990. Medical futility: its meaning and ethical implications. *Annals of Internal Medicine* 112:949–54.

Schwartz, Robert L. 1990. Euthanasia and the right to die: Nancy Cruzan and New Mexico. *New Mexico Law Review* 20: 675–695.

Scott, Vernon. 1979. Carol Burnett makes war on drugs. *Chicago Tribune*, October 28.

Sommerville, A. 2005. Changes in BMA policy on assisted dying. *British Medical Journal* 331: 686–688. https://www.ncbi.nlm.nih.gov/pmc/articles/PMC1226256/.

Speakman, Ray. 1989, 1993. Introduction and notes. In *Whose Life Is it, Anyway?*, Brian Clark, v-xi. Heinemann Plays series, Heinemann Educational Publishers, London: Oxford.

Sullivan, Joseph. 1976. Quinlan condition called irreversible by ethics unit. *New York Times*, June 11.   http://www.nytimes.com/1976/06/11/archives/quinlan-condition-called-irreversible-by-ethics-unit.html?_r=0.

The Broadway League. 1980. Cast of Whose Life Is It, Anyway? https://www.ibdb.com/broadway-production/whose-life-is-it-anyway-3686. Accessed 22 Feb 2018.

The National Commission for the Protection of Human Subjects of Biomedical and Behavioral Research. 1979. The Belmont Report: Ethical Principles and Guidelines for the Protection of Human Subjects of Research. Washington: Department of Health and Human Services.

United States Congress. S.1883—94th Congress (1975–1976). https://www.congress.gov/bill/94th-congress/senate-bill/1883.

Voluntary Euthanasia Society. 1971. Doctors and Euthanasia. (Report) (Note: these reports are referenced here: http://journals.sagepub.com/doi/pdf/10.1177/002581727304100103.).

Wilkinson v. Vesey, 110 R.I. 606, 624 (1972).

Closure and Family Systems: *My Life* (1993)

3

The film, *My Life*, is based on an original screenplay by Bruce Joel Rubin about Bob Jones (born Bob Ivanovich), who is dying from metastatic renal cell carcinoma in his late 30s. It is not a good time to have a terminal illness, as his wife is pregnant. Bob makes a series of videos about his life for his unborn son so he will not grow up without knowing his father. At the same time, Bob, born and raised in a lower middle class family and neighborhood in Detroit, does not get along with his own parents, and now lives in Los Angeles with his wife. Ultimately, the film deals with closure between terminally ill patients and their family members. This film is a "Good Death" film that lies in contrast to the "bad death" in *Wit*, discussed in Chap. 1. The clinical ethics issues in this film revolve around cancer as a transformative experience; caring for the psychosocial and spiritual needs of the dying so they can make choices that help them have a better death. The film also highlights how dying at home with hospice support can be a much better end of life experience for patients and their families. In this film, it can be argued that Bob Jones is happiest at the end of his life because he finally learns how to live well when confronting death. This is not an unrealistic or maudlin story, and will resonate with those who may work in spiritual care. The film is an example of patients able to make life choices—right to the end—which can still change their lives for the better, even if at the end of life. The critical midwife in his spiritual awakening is Dr. Ho—a Traditional Chinese Medicine (TCM) practitioner that Bob begins to see for his cancer treatment.

To be frank, this is a "lighter fare" death and dying film that endures still as a great teaching film because it does a nice job of portraying the "good death". White millennials and beyond, who may come from relatively sheltered backgrounds, may find the mood and tone of this film more digestible, although most will probably become emotional by the end—particularly fathers or expectant fathers. For multi-cultural classrooms, this film, at its core, is about the importance of closure and family, but when teaching, it will be important to discuss the character within the demographic of "white male privilege"—which doesn't make it unrealistic, but requires *cultural competency* when teaching it (see under Social Location).

© Springer International Publishing AG, part of Springer Nature 2018
M. S. Rosenthal, *Clinical Ethics on Film*,
https://doi.org/10.1007/978-3-319-90374-3_3

The Production of *My Life*

Bruce Joel Rubin, who wrote, produced and made his directorial debut with *My Life*, had previously written the screenplay for the film, *Ghost* (1990), which was a tremendous success, and nominated for several academy awards. (As it happened, Patrick Swayze, who starred as the Ghost, died young, too, of pancreatic cancer.) Rubin was working on his next script, and got the idea for *My Life* when he was sick one night with terrible gastrointestinal pain, and wondered if he had cancer. (He did not; and it was a minor stomach upset that passed.) He states (Pristin 1993):

> I lay there in bed, trying to deal with the fact that I was dying. I thought about my young children, and said, 'If I were to die, what would they know about me?' I decided that a video of me just talking about my life would be a gift that I could leave to them…And so, as I lay there trying to imagine this video, what started coming through was 'This is a great idea for a movie.'

Rubin's success with *Ghost* enabled a quick "green-lighting" of *My Life* (Shehar 2012), which premiered November 12, 1993, just in time for Thanksgiving, when family gatherings are on moviegoers' minds. The film outperformed other films premiering that weekend (Fox 1993).

A "Clintonian" Screenplay

It's important to highlight that this storyline is "Clintonian" in its appeal to audiences of its day. In 1993, a film about a dying father filming a series of videos for his unborn son may have had special significance for newly inaugurated President Bill Clinton (born 1946), who, like Bob Jones, goes through life under an assumed name. William Jefferson Clinton was not born a Clinton; he had assumed his stepfather's name, when his mother remarried when he was 4. He was born William Jefferson Blythe, III, and never knew William Jefferson Blythe, II—his father, a travelling salesman who was killed in a car crash three months before he was born. Bill Clinton sailed to victory from humble, working class roots, and indeed made clear that he never knew his father. In his own book, also called My Life (Clinton 2004), he stresses that not knowing his biological father had a negative impact on his emotional and personal life. When Rubin is developing his screenplay during the 1992 Presidential campaign, he is surely struck by the fact that Clinton was born to a widow in Hope, Arkansas. Hope was on everyone's mind when Rubin is writing the script. "I still believe in a place called Hope" was the mantra of Clinton's campaign. In the film, when Bob's oncologist leaves him hopeless, he storms back into his office unannounced and lambasts him for taking his hope away; the character makes clear that hope is the only thing he has. (It is not known if the title of Bill Clinton's autobiography is a nod to the film).

Mr./Dr. Ho

In the film, the character of the TCM practitioner is called "Mr. Ho", but should probably be referred to in teaching the film today as *Dr. Ho*, as he is supposed to be a credentialed TCM practitioner in the film, and as such, would be considered a doctor in that healing system.

When we first see Dr. Ho in the film, the viewer is also benefiting from the session, and, in some sense, is having a TCM therapy session vicariously. Haing Ngor, who plays Dr. Ho, in the film, does not seem to be "acting" but really engaged in "healing". Ngor's screen presence conveys a wisdom that is beyond a performance, and there's a reason: Ngor trained as a surgeon and gynecologist in Cambodia. When Pol Pot's Khmer Rouge ultimately seized control of the country in 1975, toward the end of the American involvement in the Vietnam War, Ngor concealed his training to avoid persecution as an intellectual and professional. Living in Phnom Penh, Cambodia, he was rounded up and imprisoned along with about 2 million other Cambodians and sent to a Khmer Rouge concentration camp along with his wife, who was pregnant. She died during childbirth because she needed a caesarean section, which he was unable to perform. After the fall of the Khmer Rouge in 1979, Ngor immigrated to the United States in 1980 after working in Thailand as a field doctor. But he was unable to get his U.S. license to practice. In 1984, he was cast in *The Killing Fields* and won the Academy Award for Best Supporting Actor in 1985. Ngor survived imprisonment by using his medical knowledge to keep himself alive by eating beetles, termites, and scorpions; he eventually crawled his way across the border to a United Nations sanctuary (Donahoe 1985; Henderson 2016). In 1988, he wrote his autobiography, Haing Ngor: A Cambodian Odyssey, describing his life under the Khmer Rouge in Cambodia (Ngor and Warner 1988). He was cast in a few other projects before being cast in *My Life*. His own life ended tragically in 1996 when he was the victim of an Asian gang murder in downtown L.A. (Henderson 2016; Tran 2010), a timeframe for the city where racial and social unrest was at its height (see further under Social Location). It is widely believed that Ngor's murder was due to random gang violence, and not premeditated.

Synopsis

In this film, there is no "life before cancer" set up, in which we meet the character prior to his cancer diagnosis and follow along with him as he gets the news. Instead, we first meet him as a child on his most embarrassing day (he tells his class that a circus will be performing in his backyard afterschool because he had "asked God" for it the night before, and thought the request would be granted). When the children show up and there is no circus, he runs to his room in shame. We then snap forward to him speaking into a video camera and telling us he has kidney cancer, not long to live, and we realize he is talking to his unborn son. We meet him and his wife, Gail (Nicole Kidman) closer to the "Acceptance" stage of death, after the couple has gone through his first three stages (anger, denial, and bargaining), and it's now part of their daily living. We learn that he has already seen many physicians, has tried and failed clinical trials, and in his final Western medicine visit, he is given a comfort care/hospice type of speech from his doctor, who reminds him that he did not respond to the last clinical trial of interleukin he was on, and so there is nothing further to try. His recommended treatment has been pain management at home, and preparedness for the symptoms of metastatic disease. Bob's wife takes him to "Mr. Ho" (who I call Dr. Ho in this chapter), a TCM practitioner played by Haing S. Ngor. Bob is a reluctant TCM patient; Dr. Ho "reads" Bob's body and emotional health accurately, and works with him on end of life goals of forgiveness and letting go of anger, but is not able to alter the course of the cancer, as he apparently can do in some other cases. Bob makes a trip home to Detroit for his brother's wedding, and the trip is rocky; he is unable to tell his family about his illness. But, he does get a "God request" granted: to live long enough to see his son born. Bob is able to spend quality time with his newborn for about 6 months. Bob continues to function reasonably well with pain control, and frequent visits to Dr. Ho, until he notices symptoms of brain metastasis; at this point, he becomes debilitated more quickly, and a hospice nurse is brought in (Queen Latifah). He finally tells his Detroit family about his illness, and they come to visit, and stay for the last few weeks of his life, as there is true forgiveness and closure. As a parting gift, his parents arrange for a circus to come to his back yard. We do see Bob's last breaths in the film, and the last scene is his son watching a video of Bob reading the Dr. Seus book, Green Eggs and Ham (1960).

The Social Location of *My Life*: White Male Yuppies and Boomers in the Early 1990s

On January 20, 1993, Bill Clinton takes the presidential oath of office and says: "The torch has been passed to a new generation born after WWII". (Recalling John F. Kennedy's inaugural line: "The torch has been passed to a new generation born in this century.") The Baby Boomers (born 1946–64) were now in charge. But what is also true is that there was one world for white boomers, and another for black, and Bill Clinton knew it; he asks Maya Angelou to deliver an inaugural poem (New York Times 1993) as a meaningful gesture to the African American community. But let's face it: he never would have won the 1992 Presidential election at that time had he been a different race, given his background.

The film, *My Life,* presents a white male character (Bob Ivanovich) that is the embodiment of what we would call today, "white male privilege". He comes from a dying demographic: a white working class, immigrant family in Detroit. He is embarrassed by his roots, and angry with his father for "working all the time". His father owns a scrap yard—a perfect small business in the car manufacturing capital of the world that defined Detroit in its heyday.

Bob has essentially estranged himself from his family. He moved to Los Angeles and reinvented himself with a new name: he calls himself "Bob Jones". Bob Jones is a classic Baby Boomer and "Yuppie" (Young Urban Professional). Bob's demographic rode a wave of success in the late 1980s, defined by a "greed is good" mentality, disposable income, and acquisition of material goods and wealth. It was during this time there was a "counter-spiritual revolution" known as the "New Age", in which many young boomers recoiled from such a vanquishing of core values, and began to seek out, in droves, spiritual nourishment and ancient wisdom which included alternative systems of healing, such as Traditional Chinese Medicine (TCM) (see further under History of Medicine) and Ayurvedic Medicine; it was in this timeframe that many Yuppies and Boomers embraced daily practices such as Yoga and Qi Gong. In fact, the more materialism that infected this demographic, the more seeking it did with respect to spiritual awareness. Rubin, himself, taught meditation, which is why he chose to create a character on the wrong "spiritual path" whose death and dying process led him to "course correct" and discover what was actually important: love and family. So although this is a "death and dying" film, it is really about spiritual awakening—to the point where some critics were annoyed. The only real relationship we see between Bob and a healthcare provider surrounds his experiences with his TCM practitioner, Dr. Ho. The film's message to its white boomer audience: Wake Up! There's much more to life than *your* life. Yet reviews about the film were critical of its New Age content: "In its relentless Hollywoodization of what in most people's lives would be an agonizing situation, this film actually makes death look like a negligible price to pay for the spiritual wealth gained by opening up and becoming a caring human being" (Turan 1993).

Some reviewers noted a trend: several films in the early 1990s were about professional white males suddenly facing illness; some suggested it was "a kind of cover for the results of the [1990] recession" in which "yuppie achievers" needed to "discover their humanity and realize that life wasn't all about condos and limos" (Pristin 1993). It was also a commentary on death from AIDS (see further).

Toxic L.A.

In the early 1990s, before the infamous L.A. riots, which occurred in 1992 in the wake of the Rodney King verdict, several "L.A. commentary" films emerged that spoke to the urban social disaster that had become Los Angeles. In 1991, the film, *Grand Canyon*, explored the troubled lives of L.A. residents—black and white. Each character, we discover, is living in "hell", and the gritty realism of what L.A. had become by the early 1990s was on full display. "This town stinks," as one of the white characters comments. If you were white and wealthy, homeless people roamed your streets and you had to pretend not to see them; helicopters were overhead day and night in black neighborhoods that might as well have been a different country, ruled by gun violence and gang warfare; carjackings were a continuous threat. In the black part of town, gang violence is chronic, and people are forced to live as though they are in a war zone, as their homes become targets of "drive by" machine-gun violence from gang assassins seeking to take-out a "high value" target from their vehicles. The film brings together the white character (Kevin Kline) and black character (Danny Glover); each uses his cultural competencies in his own part of town to help the other as they discover a great friendship. *Boyz in the Hood* (1991) was about the black youth experience in South Central L.A. that framed gangbanging as a way of life in a hopeless community of distributive injustice.

The next year, the horrendous L.A. riots occurred (April 29-May 4, 1992), which were a prelude to the extreme racial overtones in the O. J. Simpson trial (1994–5). The history of race in Los Angeles is an important discussion when teaching this film because you can't discuss a "white guy" dying, without discussing what is happening a few blocks away from his nice house where he is receiving hospice care. The only black person in the film is Queen Latifah, who plays the Hospice Nurse, but she is not a commentary on black life in L.A., and is not really a three-dimensional character in the film (Ebert 1993).

Ultimately, there were reams of "Bob Jones-like" white males in Los Angeles living with blinders on, who never went to the "bad parts" of town. L.A. becomes a microcosm for American urban life at that time: two Americas—one white; one black. *It's not good, but it's still real.* In this case, we are in the white part of town; Bob Jones is a dying white guy. There were plenty of other dying "white guys" in Los Angeles at the same time: but they were dying of AIDS (see further).

In 1993, the same year *My Life* premiered, two other L.A. commentary films debuted that focused on the white demographic experiences in L.A.: *Short Cuts* (1993), a Robert Altman film, which adapts nine Raymond Carver short stories into the "toxic L.A." genre. *Short Cuts* (in development by 1989) exposes L.A. life as a gritty hellhole for its white characters, who span genders and income brackets, with a focus on "White Trash" in L.A. *Falling Down* (1993) is about white male rage: a professional, "square" white male (Michael Douglas) "loses it" in a traffic jam and becomes unleashed with his pent-up rage.

From a cultural competency standpoint, it's important to note that the seeking of "spirituality" and "New Age" remedies in toxic 1990s L.A. was in reaction to what seemed to be a "wicked problem" of racial and income injustice in L.A. By the time *My Life* is widely released in 1994, L.A. exploded again in racial tensions when black and white cultures became equally entangled and invested in the O. J. Simpson murder case, a topic that was recently examined fully in a several hours-long ESPN 2016 documentary titled *O.J.: Made in America* (Edelman 2016), which won an academy award. So while race is not a spoken subject in *My Life*, it cannot be an unspoken subject when teaching a film that takes place in early 1990s L.A.

The Asian American Experience in 1990s L.A.

Culturally, the Asian-American experience in this racially mixed city could be explored through a discussion of the character of Dr. Ho, and the actor who played him, Dr. Haing S. Ngor, who was killed by Asian gang members in downtown L.A. in 1996. Of note, the L.A. riots of 1992 pitted African Americans against Asian

Americans (Korean retailers, in particular). Prior to the Rodney King beating, a March, 1991 altercation between a Korean grocery store owner and a 15 year-old African American customer ended in murder, when the grocer shot the customer over a misunderstanding involving orange juice. A jury found the grocer guilty, but a (white) judge sentenced her to primarily probation and "community service", which enraged the black community (Wilkinson and Clifford 1991; Edelman 2016). During the 1992 riots, Korean shops were particularly targeted (Singleton et al. 2017; Lindsay and Martin 2017). By 1994, Asian Americans were next (unfairly) cast as "quintessential" unbiased players in the L.A. justice system (as jurors, witnesses or jurists) because they were thought to be the "honest brokers" of black versus white tensions. However, this did not work well in the courtroom of Judge Lance Ito, the judge selected for the Simpson case, who, while of Japanese descent, was married to a white woman (like O.J.), and criticized for basking in the media attention that became a conflict of commitment for him. Many had assigned "higher order" virtues to Asian-Americans as the moral arbiters of racial order in L.A., but Asian Americans were also living in "hell" in early 1990s L.A.

In teaching more about Ngor's life, students should see that gang violence affected Asian Americans as well, but it was a more occult phenomenon that received little, if any, media coverage. One could spend time discussing more about Dr. Ngor's life and experiences as a bridge to more on this topic. For example, depending on the rest of the curriculum when showing this film, one could bridge this film to a pediatric ethics "classic"—a popular cultural bioethics book, The Spirit Catches You and You Fall Down: A Hmong Child, Her American Doctors, and the Collision of Two Cultures (Fadiman 1997). This non-fiction book is about the Hmong population (the indigenous population of Laos) in California, which became prevalent after the Vietnam War. The book tracks the failure of Western practitioners to properly understand and care for a Hmong child with epilepsy.

At minimum, however, what needs to be emphasized when discussing the character of Dr. Ho is realistic access to TCM, as it is still part of the "White Death" experience when viewed from an access perspective. Any alternative medicine practitioner, including TCM, is only accessible to people with disposable income, since no alternative healing is covered by any health insurance plan (even now). This is explored further on (see History of Medicine section).

Detroit: A Dying City

My Life features Detroit prominently as the hometown of its dying character. Rubin also grew up in Detroit, so its role in the script may be partly autobiographical. However, in a death and dying film, the city itself was notable by 1990 to be on "life support." Detroit, like L.A., had many racial problems, and in 1967, one of the largest riots in U.S. history broke out in Detroit. Its white middle-class population began fleeing the city (Stone 2015). In 1990, the *New York Times* featured an article entitled "The Tragedy of Detroit" which stated the following (Chafetz 1990):

Detroit is a city of one- and two-story homes, most of them built on narrow lots. During the last 30 years, the city has lost almost half its population, and there are entire blocks where all but one or two houses are boarded up and vacant. Some parts of the city look like pasture land. Flames raced through the brush and into abandoned buildings. The gawkers cheered the firemen and jostled one another happily. The geography of Detroit has not changed since my childhood. It is the demography that is different. In 1960, there were 1,670,000 people in the city, about 70 percent of them white. Poles and Italians lived in neat, boxlike homes on the east side. Jews and WASP's inhabited more substantial brick houses on the other side of Woodward Avenue. Blacks, who made up less than a third of the population, were crowded mostly into small neighborhoods downtown, near the river.

In the film, it's not entirely clear where Bob's parents were born, but their accents suggest they are Eastern European. We can assume they, too, fled the city for the suburbs along with other white ethnic groups (Chafetz 1990):

The White abandonment of Detroit, coupled with the collapse of the auto economy, has left the city with a diminished tax base and a set of horrific social problems. Among the nation's major cities, Detroit was at or near the top in unemployment, poverty per capita and infant mortality throughout the 1980s.

Like L.A., the African American inhabitants were not the merchants of Detroit; by 1990, Detroit became home to the largest Arab population in the U.S., who replaced all white ethnicities as the local shopkeepers. Essentially, the Arab population was to Detroit what the Korean population was to L.A. Thus, a dying man goes home to visit his "dying city", which may have been an unintentional symbol, but should be discussed, as he seems to have an aversion to going home.

We first see Detroit in 1963, when Bob is a child, and in a few more childhood flashbacks. We next see it at the wedding; the parents have presumably moved to the suburbs (probably Macamb County, which is where many European families fled from Detroit). Bob visits his old backyard where his G.I. Joe doll is still hidden in a brick, and he is then discovered by the current African American family living there, who invite him in so he can see the house again. It must be stated that if it were the reverse: an African American male discovered in a white owner's yard, he would surely be arrested or shot.

The story of Detroit gets worse; the last major supermarket chain store closed in 2007 (NPR 2007; Harrison 2009) making the city a huge food desert for many years. The city also declared bankruptcy in 2013.

Ultimately, it's notable that the two cities the filmmaker chose for the death of his white male character's dwellings were considered the most racially problematic in the country in this time frame.

White Males Dying of AIDS

Movies about white males and illness were part of the "Zeigeist" of the early 1990s, as deaths from AIDS began to have a major cultural impact. An article that appeared in the *Los Angeles Times* in 1993 specifically commented on the number of films being made with the theme of white men facing illness or death (Pristin

1993) which included *The Doctor* (1991), which I discuss in <u>Healthcare Ethics on Film</u>, and *Regarding Henry* (1991), about a gun violence victim who sustains a brain injury. Pristin notes:

> In the last 10 years the movie community has known a lot of people who died from AIDS, and a lot of people under the age of 40 have had to encounter more death than most people would have at their age...That has introduced them to the subject of mourning, and it's not too surprising to see them try to play it out on the screen.... [F]ilm critics and other Hollywood observers attribute the rekindled interest in dying primarily to the disproportionate effect of AIDS on the creative community.

In 1993, the critically acclaimed film, *Philadelphia,* also premieres. The film is about a lawyer dying of AIDS, who is suing his law firm for being dismissed due to discrimination; it won Tom Hanks the Best Actor Academy Award. (I explore the subject of AIDS-related deaths in <u>Healthcare Ethics on Film</u>.)

The Role of Technology in Death and Dying

By 1993, the Sony camcorder (a device that was both a video camera and recorder in one) had become an indispensable tool in most middle-class households, capturing major life events. First introduced in 1983, it debuted just as the Baby Boomers began to have their own families. In 1993, the camcorder in use in the film was bulkier than the digital video standard introduced in 1995 that reigned well into the 2000s. The use of video in documenting memoir, and interviewing elder family members was a common phenomenon at this time. Self-documentary would only become more popular in the next few decades as technologies improved, especially when the iPhone was introduced in 2007. The idea of pre-recording a series of videos about yourself to talk an unborn child you may not meet was a natural concept for a filmmaker, such as Rubin, but an innovative idea to its audience at the time. *My Life* may have sparked an occult trend that no one has properly documented. Several companies emerged in the early 1990s that offered professional family memoir taping and production—sometimes making use of professional journalists or documentarians. It was usually to tape family elders.

In the final scene of *My Life*, where Bob's son (Brian), now at least a year old, is watching him read Dr. Seus' classic 1960 book, <u>Green Eggs and Ham</u> and points to the television to say "Dada"—it is clear that a relationship is forming. The Baby Boomer generation was also known as the first "television generation" that grew up on children's programming. Daily doses of *Mr. Rogers Neighborhood* (1966–2008), for example, provided children with a positive male role model that many considered an important part of their childhood (PBS 2009). A popular reading show when *My Life* was made was *Reading Rainbow* (1983–2006), which presented a different book and theme each episode featuring many celebrities reading the books. In Brian's case, watching "Dada" on television everyday, who filmed a wide variety of content just for him, may wind up representing more time spent with his father than what many children grew up with. Statistics of that time revealed that children spent more time watching television than quality time with their families (Rosenthal 2004). In this case, Brian's TV time is his quality time with Dad. It's worth exploring, when teaching, current perspectives from students about the social relationships they forge using "screen-based" technologies.

When exploring self-documentary, it's important to frame it as having its roots in cinema vérité (meaning "truth on film"), a style of documentary film-making that was "unedited" and pioneered in the U.S. by Richard Leacock, D. A. Pennebaker, and Albert and David Maysles (Axmaker 2015), known for the cult classic, *Grey Gardens* (1975). The cinema vérité style was adapted for "reality television", and was the stylistic bridge to social media, which has become the dominant form of communication for interpersonal relationships (with its own set of ethical, legal, and social implications). Citizen journalism was another outgrowth of the widespread use of camcorders. In fact, one of the first "viral videos" was taken on March 3, 1991, of the Rodney King beating (see earlier); the 1992 L.A. riots involved many citizen journalism videos shot with individual camcorders.

In 1993, the same year *My Life* is released, the documentary, *Silver Lake Life* (1993) was a day-by-day "reality series" that documented the final days of a man dying of AIDS; the entire film is shot with a hand-held camcorder.

The History of Medicine Context

There are three history of medicine contexts to explore with this film: treatment of renal cell carcinoma in a timeframe prior to targeted therapies such Sorafenib, which is a protein kinase inhibitor; the status of hospice care in this timeframe; and finally, the emerging popularity of Traditional Chinese Medicine (TCM) by U.S. patients during this timeframe, and its role in end of life care.

Renal Cell Carcinoma

Bob Jones has metastatic renal cell carcinoma, the most common form of kidney cancer, but which is relatively uncommon, compared to other cancers. In the United States, renal cell carcinoma (RCC) accounts for 2.6% of all cancers, but when it does occur, it is 60% more likely in men than women (Mandal 2014; Godley and Ataga 2000; El Fettou et al. 2002). In the 1990s, there were very few options for advanced renal cell carcinoma, and clinical trials were available with the drug interleukin, which is referenced in the film (Amin and White 2014; Rosenberg 2007; Schwartz et al. 2002; ASCO 2017). (Bob has tried, and failed, on a previous trial of interleukin.) Renal cell carcinoma can spread quickly, and in most cases, there are no early warning signs, so patients may present when it is in an advanced stage. Usually the signs of this type of cancer are not specific until the cancer progresses to a more advanced stage. In about 25% of cases, the cancer is diagnosed in its advanced stage. Based on today's estimates, once renal cell carcinoma has spread to other organ systems in the body, the 5-year survival plummets to about

5-15%, which is due to the availability of protein kinase inhibitors. (Amin and White 2014; Turhal 2002; Walid and Johnston 2009). But in 1993, there were no treatments at all, and Bob's prognosis is accurate; when we first meet him, it has already spread to his lungs and he now has considerable pain, signaling to us that it is in a very advanced stage now. In the film, we learn that it metastasizes to the brain, which even today, would carry dismal survival rates, but with a protein kinase inhibitor, could extend life for at least a year about 50% of the time (ASCO 2017; Santoni et al. 2012). Without such therapies available in 1993, there was no treatment. However, it would be worth discussing in class what a difference one year would make when you want to spend time with a young child. A discussion today would entail access to protein kinase inhibitors, which are a very expensive drug, costing thousands of dollars a year.

Generally, renal cell carcinoma is diagnosed in men over 60, but Bob Jones is young—in his late 30s, if we assume he is about 7 when we first meet him in 1963 Detroit. We have hints that he has a "rare" form of this cancer, and that's because sporadic renal cell carcinoma is rare in younger men, accounting only for about 3.4% of cases (Mandal 2014). For these reasons, even in cases where there are more obvious symptoms, it would be harder for practitioners to suspect this type of tumor in a younger male (El Fettou et al. 2002).

By 1992, the Food and Drug Administration approved interleukin-2 (IL-2), for advanced renal cell carcinoma (ASCO 2017), and it remains the standard of care even now, as it is the only drug that has demonstrated a cure in some patients. But fewer than 20% of patients respond to IL-2 (ASCO 2017). It is therefore not unusual that Bob has not responded. He does express interest in doing another trial, and is told he is not a candidate, which is what motivates him to seek out alternative therapy with Dr. Ho. It's important to note, too, that interleukin 2 does not have alopecia as a side-effect, but carries many other toxicities. That's why it's perfectly realistic that Bob has not lost his hair when we meet him in the film; however, it also helps to explain why his physician is reluctant to put him on another trial, which does not offer benefit. In 2002, it was noted that: "Because of intolerable side effects, most patients do not receive 100% of the planned dosing in a full cycle or course of high-dose IL-2. Furthermore, most IL-2 administration and toxicity management guidelines recommend withholding therapy, not reducing the dose, in patients who experience various toxicities" (Schwartz et al. 2002). This explains why Bob was told frankly by his physician that he would not be a candidate for a second trial of IL-2.

The State of Hospice Care: 1993

Hospice care was just beginning to be utilized by large segments of the population in 1993. From the root word, "hospitality", the term "hospice" was used as early as the medieval period to mean a place of "rest for the weary or ill"—often reserved for travelers after very long journeys, where it used to take weeks to travel to the same places that are now a one hour plane ride. The idea of a hospice for the dying, was first promoted by a female physician in England, Dame Cicely Saunders, who started to work with dying patients individually using a hospice model (symptom/pain control only), and created the first modern hospice in 1968 (St. Christopher's Hospice) in a suburb of London (NHPCO 2016; Connor 1998). However, it should be noted that the first known patient to receive in-patient hospice care was Joseph Merrick in the 1850s (see Chap. 8, on *The Elephant Man*).

The idea of hospice in the U.S. was also suggested by Saunders in 1963, at a lecture she gave at Yale University to a multi-disciplinary healthcare audience (nurses, medical students, social workers, etc.). In the lecture, she showed photographs of dying patients "before" and "after" pain control with their families, and the results were appealing. Two years later (1965), the Dean of Yale's nursing school, Florence Wald, invited her to be a visiting professor, and then by 1968, Saunders opened the first official Hospice location in London. Then the "bombshell" publication by Elizabeth Kubler-Ross, On Death and Dying (1969), was published, which consisted of over 500 qualitative interviews with dying patients, which made it clear that home hospice was desirable for patients; Ross testifies in 1972 regarding death with dignity to the U.S. Senate Special Committee on Aging and said this (NHPCO 2016):

> We live in a very particular death-denying society. We isolate both the dying and the old, and it serves a purpose. They are reminders of our own mortality. We should not institutionalize people. We can give families more help with home care and visiting nurses, giving the families and the patients the spiritual, emotional, and financial help in order to facilitate the final care at home.

By 1974, Wald, along with two pediatricians and a chaplain, founded Connecticut Hospice in Branford, Connecticut, and the first hospice legislation is introduced to provide federal funds for hospice programs, but doesn't pass. Lobbying for federal government support ensues between 1978–1980, when the W. K. Kellogg Foundation funds research by the Joint Commission on Accreditation of Hospitals (JCAHO) to develop standards for hospice accreditation. Medicare added hospice services to its coverage in 1982 (Connor 1998). On September 13, 1982, President Reagan announces "National Hospice Week" (Federal Register 1982) in "Proclamation 4966—National Hospice Week, 1982" in which he states:

> The hospice concept is rapidly becoming a part of the Nation's health care system... Hospice provides a humanitarian way for a terminally ill patient to approach death with dignity, in relative comfort in a supportive atmosphere, and surrounded by family members. Its most important element is concern for patients and their families. Hospice advocates personal care and concern, living comfortably until death, the absence of pain, maintenance of personal control, and the close fellowship of the family unit.

Thus, Medicare added hospice services to its coverage in 1982, and by 1984, accreditation standards emerge. In 1986, Congress votes to fund a permanent Medicare Hospice Benefit, and states are given the option to include hospice in their Medicaid programs, but by 1989, only 35% of established hospices are Medicare-certified, so many patients do not qualify; many hospice services are billed out-of-pocket to patients. This slowly begins to change; by 1992, Congress calls for another feasibility study of hospice, and by 1993, when the film is made,

President Clinton's promising *Health Security Act* bill (which dies in Congress in 1994), includes a nationally guaranteed benefit (NHPCO 2016), which likely inspired Rubin to show what hospice care looks like for a dying man and his family. When this film premieres, hospice care becomes a provision for some patients and not others, depending upon health insurance plans, which are highly heterogeneous. We are not privy, in the film, to Bob's insurance coverage, but as a white male who owns a profitable business, we can assume that he probably has a good health insurance plan; hospice may be covered by his plan when there is no other treatment available. Prior to federal funding for Hospice, the first hospital-based palliative care programs began at the Cleveland Clinic, the Medical College of Wisconsin and a handful of other U.S. institutions, and so it's possible that Bob's institution has such a program in which home-based hospice is provided. Regardless of our speculation as to how Bob is paying for Hospice, by 1993, he is among thousands of dying Americans who are receiving hospice care in either their hospital institutions, or at home. Patients who are told they have less than six months to live qualify for Hospice, which can create other problems if they live longer than expected—in which case—they are "re-certified" for hospice care. By 1998, there were 3200 hospices either in operation or under development throughout the United States (Plocher and Metzger 2001). The World Health Organization currently defines palliative care as: "an approach that improves the quality of life of patients and their families facing the problems associated with life-threatening illness, through the prevention and relief of suffering by means of early identification and impeccable assessment and treatment of pain and other problems, physical, psychosocial and spiritual" (WHO 2017).

Traditional Chinese Medicine (TCM)

It's critical to note that the most important healthcare relationship highlighted in this film is not between Bob and his allopathic physicians—physicians who are not given a name in the film—but between Bob and Dr. Ho, the (presumed) TCM practitioner he seeks out, who is described by one film reviewer as "a mysterious master of undefined Asian medicine" (Turan 1993). First, as noted above, the character of the TCM practitioner is called "Mr. Ho"—which was a cultural competency error made by the 1993 screenwriter, Rubin. Ironically, "Mr." Ho is played by an allopathic physician from Cambodia, discussed earlier. It is thus appropriate today to acknowledge, when teaching, that we generally do refer to alternative medicine practitioners as "doctor" if they complete training within their own traditions; they just may not be designated as Medical Doctor (M.D.).

When *My Life* premieres, TCM was a burgeoning trend—particularly in California, which was always a trend-setting state with respect to where the cultural winds were blowing. In 1976, California was the first state to establish an Acupuncture Board and became the first state licensing professional acupuncturists, a basic practice of TCM. By the early 1990s, there was a sharp rise in the growth and popularity of alternative medicine in general, and TCM in particular. (Eisenberg et al.

1998; Whorton 2002; Coulter and Willis 2004). (This late twentieth century trend lies in sharp contrast to the treatment of the Chinese by California in the nineteenth century, and the terrible history of the *Chinese Exclusion Act* of 1882.)

When teaching, it may even be useful to invite a TCM practitioner to class to demonstrate some of the fundamental principles of TCM. As an overview, TCM has evolved over thousands of years, and did originate in ancient China; it involves herbal medicine; mind-body practices, such as acupuncture, Chinese Therapeutic Massage, dietary therapies, and tai chi, but it involves understanding the body based on the qi (pronounced "chi"), considered the life-flowing energy in the body. A TCM practitioner could engage a classroom in some basic qi gong (pronounced "chi-kong") exercises, for example, to explain what is meant by qi. Qi gong combines specific movements or postures, coordinated breathing, and mental focus. It appears in the film that Dr. Ho is primarily using Chinese Therapeutic Massage (called "tui na") to work with Bob, who is definitely noticing changes and benefit from Dr. Ho's sessions. Dr. Ho ultimately helps to explain to Bob a different perspective on his cancer as an "emotional blockage" that is very deep. While this would cause allopathic practitioners to balk, it is a realistic presentation of the types of encounters patients have with TCM practitioners. Given that there is no treatment from allopathic systems of healing, Dr. Ho's work with Bob on his emotional health—mainly anger and forgiveness—is vital to enable proper closure with his family at the end of his life.

But the sessions with Dr. Ho are not just for Bob—*but for the audience*. Bob's feelings about his family resonates with many of us, and many of the feelings that *My Life* evokes results from the viewer's identifying with Bob's experience. Dr. Ho helps the viewers "forgive" and understand the destructiveness of their own anger with family members, too. In the end, when we envision our own deaths, we want what Bob has—closure with our family members based on unconditional acceptance and love, and the realization that anger is not helpful at the end of life.

In 1993, it was not known how many patients utilized TCM in the U.S. (Sampson 1995), but by 1997, when statistics became available, roughly 10,000 TCM practitioners saw one million patients per year (Eisenberg et al. 1998). Between 1997 and 2007, the National Health Interview Survey (NHIS) tracked 2.3 million Americans practicing tai chi and 600,000 practicing qi gong in the previous year; many were using TCM as a preventative health model. Although I do not provide a substantive overview of the philosophy of the TCM model here, it may be prudent to provide materials to students when teaching this film that are sufficiently detailed about the basic tenets of TCM. Credentialing of TCM practitioners varies from state to state in the U.S.

Clinical Ethics Issues

This is not a film about patient autonomy, competency, or patient's rights. At its core, this is a film about "the good death". This film highlights why hospice care can be the most beneficent option at the end of life. *My Life* demonstrates that hospice in the home permits quality of life at the end of life so that necessary closure with family members can occur (Connor 1998). This film punctuates that one patient's end of life experience affects the whole family, or *family system*. Hospice care is a family affair; it's as much about quality of life for the dying as it is quality of life for the caregivers and those who are left behind. As Mehta and Cohen (2009) note:

> To deliver the best care, we need to understand that the patient is immersed in a context called family, an interactive system. Death and dying should be perceived as a family event that likely throws the family out of balance and requires adjustment of all family members to the new family reality. ... The upcoming loss of a family member presents an enormous challenge, as the family must try to find new balance during the illness and then in the absence of an integral member. Furthermore, caring for a palliative cancer patient in the home also forces the family to reorganize as they learn the intricacies of caring for the dying.

Ultimately, which is germane to the film: "The experience of death and dying cannot be addressed in isolation, restricted to the patient alone. Palliative patients often face crises in symptom management, recurrent visits to the emergency room, and changes of roles within the family" (Haddad 2004). Thus, it is not possible for any patient to have a "good death" experience if s/he is disconnected from his or her family unit, even when a patient's decisions may not necessarily mirror the same decisions the family may want. It is also not possible for a family to have a "good death" experience when they are isolated or disconnected emotionally from their dying loved one. For these reasons, the concept of "comfort care"—keeping patients comfortable without painful or invasive treatments—permits them to heal emotional wounds with family members through quality time and closure (Connor 1998). Ideally, this film should coincide with materials on Family Systems Theory, a framework for delivering health care that takes into the consideration how the family unit is affected, and that the patient is *part* of his or her family unit (Kerr 2000). Even in cases with the unbefriended or unrepresented patient, as in *Wit* (Chap. 1), the absence of family can be mitigated by ensuring that a psychosocial support structure is part of the care plan.

In *My Life*, Bob has distanced himself from his parental home and roots, and has focused on his developing family with Gail and his son; Bob lives long enough to spend at least the first 6–9 months with his son. (The film does not make this time line clear, but we can guesstimate.) But Bob learns through his dying process, that he needs to bring *all* the family members together to have true happiness at the end of life, something Dr. Ho helps him to understand. When Bob enters the final stages of his illness, (his cancer has now spread to the brain) and he begins to become very symptomatic and deteriorate, it can be argued that it is also the richest time. He finally discloses to his Detroit family (parents, brother and sister-in-law) that he is dying. On his earlier visit, there is finger-pointing surrounding whose fault it is for "estrangement"; his parents argue that Bob's mother will not fly, and so his living in Los Angeles makes it impossible for them to visit. But when he finally tells them he is dying, and his whole family flies to L.A. to be with him, his mother literally "rises" to the occasion, and so does Bob. The film demonstrates that Bob's last weeks are filled with joy for him; he achieves a level of communication with his family members that would likely not have occurred had he lived another 40 years. (For example, we could easily envision Bob, in an alternate universe, getting a call that his father had died, and going to the funeral completely disconnected without closure.)

Family Systems Theory, also known as "Bowen Theory" (Kerr 2000; Family Systems Theory 2017) is not universally taught to healthcare providers, and it is sometimes counterintuitive when we emphasize patient autonomy, and that family members have to honor patient preferences when they disagree. But understanding family systems theory can also help to explain why some patients may choose options that "please" family members when it may not be in their own best interests, such as aggressive and invasive therapies that prolong their dying. (Some patients have higher-order preferences to please family members and try to "keep going" for example, which can cause considerable moral distress for healthcare providers.)

Family Systems Theory is a framework to guide the discussion of *My Life*. This theoretical model focuses on the family as a dynamic, interactive system where illness in one family member affects all family members because the family unit is interdependent. Even in cases where a family member is not conscious (as in the Quinlan or Cruzan cases discussed in the previous chapter), the family is profoundly affected by the lack of closure, and it is the family system that goes to court. Thus, a major concept within Family Systems theory is "holism"—that "the family as a whole is greater than the sum of its parts" (Mehta and Cohen 2009). Patients' preferences and decisions are best understood in relation to the family. Some of the most complex clinical ethics cases, in fact, are very straightforward medically, but are made complex because of a family system, where there may be a number of competing preferences that are affecting patient decision-making, or even confusing optimal goals of care. Some families thrive on conflict; some families work hard at closure. Family Systems Theory (Family Systems Theory 2017) advises practitioners to see the patient as part of an interactive family unit in which illness and decisions affect all the family members. This can be critical in terms of discharge planning as caregiving may profoundly affect family members and the family system. Often, family meetings surrounding goals of care will deviate into complex discussions about family finances, and many other psychosocial issues that healthcare providers may think are extraneous to the patient's medical needs. "When a family member is diagnosed with a terminal illness, the family has to reorganize itself. The individual members and the unit as a whole may not function as they previously had." (Mehta and Cohen 2009).

The circus that Bob wishes for as a child, which his parents arrange as a surprise when he is dying, can be seen as a symbol of Bob's own perception of his "family system." A circus is, by definition, a traveling group of performers with different skills and talents, who cannot be a "circus" in isolation, but are all parts of a whole. For this reason, circus performers think of themselves as a "family". For Bob, who feels let down by his workaholic father as a child, he wishes for a "circus" in his backyard because, as we may reasonably extrapolate, he is wishing for a "happy" family unit at his house. When he tells his friends to come to his house to see the circus, and finds it is not there, it affects him lifelong as a bitter memory. But at the end of life, the family unit has healed, and is happy, and thus, the circus is finally at his house, and we can further extrapolate that his parents seem to understand what his childhood wish was all about.

When applying *My Life* to bioethics core principles, we can understand hospice and a palliative care approach as the clear medical option that maximizes benefits for Bob, and minimizes harms. As his cancer follows its natural course of progression, and his symptoms are managed, he is able to be home where he is comfortable, surrounded by his family, and feel his family's unconditional love and acceptance. In this case, it's all about the psychosocial needs, which are, arguably, more important at the end of life than prolonging death with invasive therapies that erode quality of life.

Conclusions

When planning *My Life* as part of a medical ethics curriculum, it can be nicely paired with *The Elephant Man* (Chap. 8) to engage in a Hospice-as-Beneficent care discussion, or *Wit* (Chap. 1), when looking for the "antithesis" to a "bad death" experience. Clearly, when we juxtapose Bob Jones' end of life experience with Vivien Bearing's, we choose Bob's over that of a clinical trial with harsh side-effects that may even hasten a natural death.

Film Stats and Trivia

- Haing S. Ngor is the first actor of Asian descent to ever win an Academy Award for Best Supporting Actor.
- In 1997, The Dr. Haing S. Ngor Foundation was founded in his honor, which raises funds for Cambodian aid. (It is currently run by Ngor's niece).

From the Theatrical Poster

Director	Bruce Joel Rubin
Producer	Hunt Lowry; Bruce Joel Rubin; Jerry Zucker
Writer	Bruce Joel Rubin
Starring	Michael Keaton; Nicole Kidman, Haing S. Ngor; Queen Latifah
Music	John Barry
Cinematography	Peter James

Editor	Richard Chew
Distributor	Columbia Pictures
Release Date	November 12, 1993
Run time	117 min

References

American Society of Clinical Oncology (ASCO). 2017. Cancer progress timeline [in] kidney cancer. http://cancerprogress.net/timeline/kidney. Accessed 23 Feb 2018.

Amin, A., and R. White. 2014. Interleukin-2 in renal cell carcinoma: A has-been or a still-viable option? *Journal of Kidney Cancer VHL* 1: 74–83. https://www.ncbi.nlm.nih.gov/pmc/articles/PMC5345537/.

Axmaker, Sean. 2015. Cinema Verite: The moment of truth. PBS.org, December 14. http://www.pbs.org/independentlens/blog/cinema-verite-the-movement-of-truth/.

Chafetz, Ze'ev. 1990. The tragedy of Detroit. *New York Times*, July 29. http://www.nytimes.com/1990/07/29/magazine/the-tragedy-of-detroit.html?pagewanted=all.

Clinton, Bill. 2004. *My Life*. New York: Alfred A. Knopf.

Connor, Stephen R. 1998. *Hospice: Practice, Pitfalls, and Promise*. Oxford: Taylor & Francis.

Coulter, Ian D., and Evan M. Willis. 2004. The rise and rise of complementary and alternative medicine: A sociological perspective. *Medical Journal of Australia* 180: 587–589.

Donahoe, Deirdre. 1985. Cambodian doctor Haing Ngor turns actor in the killing fields, and relives his grisly past. *People*, February 4. http://people.com/archive/cambodian-doctor-haing-ngor-turns-actor-in-the-killing-fields-and-relives-his-grisly-past-vol-23-no-5/.

Ebert, Roger. 1993. My Life. Roger Ebert.com, November 12. http://www.rogerebert.com/reviews/my-life-1993.

Edelman, Ezra. 2016. *O.J. Made in America*. ESPN Films and Laylow Films, released January 22.

Eisenberg, David, Roger B. Davis, Susan L. Ettner, Scott Appel, Sonja Wilkey, Maria Van Rompay, and Ronald C. Kessler. 1998. Trends in alternative medicine use in the United States, 1990–1997: Results of a follow-up national survey. *Journal of the American Medical Association* 280: 1569–1575.

El Fettou, H.A., E.E. Cherulo, M. El Jack, Y. Al Maslamani, and A.C. Novick. 2002. Sporadic renal cell carcinoma in young adults: Presentation, treatment, and outcome. *Urology* 60: 806–810. http://www.goldjournal.net/article/S0090-4295(02)01884-8/fulltext?cc=y=.

Fadiman, Anne. 1997. *The Spirit Catches You and You Fall Down: A Hmong Child, Her American Doctors, and the Collision of Two Cultures*. New York: Farrar, Straus and Giroux.

Family Systems Theory. 2017. http://www.familysystemstheory.com. Accessed 19 July 2017.

Federal Register. Proclamation 4966. September 13, 1982. https://www.reaganlibrary.archives.gov/archives/speeches/1982/91382h.htm. Accessed 19 July 2017.

Fox, David J. 1993. Swords duel 'Carlito'. *Los Angeles Times*, November 15. http://articles.latimes.com/1993-11-15/entertainment/ca-57199_1_box-office.

Godley, Paul A., and Kenneth.I. Ataga. 2000. Renal cell carcinoma. *Current Opinion in Oncology* 12: 260–4.

Haddad, A. 2004. Ethics in Action: End-of-life decisions: The family's role. *Modern Medicine*. http://www.modernmedicine.com/modern-medicine/content/ethics-action-end-life-decisions-familys-role.

Harrison, Sheena. 2009. A city without chain grocery stores. *CNN Money*, July 22. http://money.cnn.com/2009/07/22/smallbusiness/detroit_grocery_stores.smb/index.htm.

Henderson, Simon. 2016. The Life and Strange Death of the Khmer Rouge Survivor Who Won an Oscar, Then Got Murdered. *Vice News*, February 25. https://www.vice.com/en_us/article/av39pg/remembering-the-khmer-rouge-survivor-who-was-killed-for-his-oscar-winning-role.

Inauguration; Maya Angelou: 'On the Pulse of Morning'.1993. *New York Times*, January 21. http://www.nytimes.com/1993/01/21/us/the-inauguration-maya-angelou-on-the-pulse-of-morning.html.

Kerr, Michael E. 2000. One family's story: A primer on Bowen Theory. The Bowen Center for the Study of the Family. http://www.thebowencenter.org. Accessed 19 Jul 2017.

Lindsay, D., and T.J. Martin. 2017. *LA 92*. National Geographic Documentaries, released April 28.

Mandal, A. 2014. Renal Cell Carcinoma Epidemiology. *Life Sciences Medical News*. http://www.news-medical.net/health/Renal-Cell-Carcinoma-Epidemiology.aspx.

Mehta, A., and S.R. Cohen. 2009. Palliative care: A need for a family systems approach. *Palliative and Supportive Care* 7: 235–243. https://www.ncbi.nlm.nih.gov/pubmed/19538807.

National Public Radio (NPR). 2007. No more supermarkets: Major grocers flee Detroit, August 3. http://www.npr.org/templates/story/story.php?storyId=12477872.

National Hospice and Palliative Care Organization. 2016. History of Hospice. https://www.nhpco.org/history-hospice-care. Accessed 19 July 2017.

Ngor, Haing, and Roger Warner. 1988. *Haing Ngor: A Cambodian Odyssey, describing his life under the Khmer Rouge in Cambodia*. New York: Macmillan.

PBS Kids and Family Communications, Inc. 2009. Dear Mr. Rogers Commemoration Page. http://pbskids.org/rogers/dearMrRogers2168.html. Accessed 19 July 2017.

Plocher, David W., and Patricia L. Metzger. 2001. *The Case Manager's Training Manual*. Sudbury, Massachussets: Jones & Bartlett Publishers.

Pristin, Terry. 1993. Death Takes This Holiday: Films about dying have normally been taboo for box-office-conscious studios. So what's with all these movies about death during this holiday movie season? Don't worry, some predict the trend will be short-lived. *Los Angeles Times*, November 21. http://articles.latimes.com/1993-11-21/entertainment/ca-59228_1_holiday-movie-season.

Rosenberg, S.A. 2007. Interleukin 2 for patients with renal cancer. *Nature Clinical Practice Oncology* 4: 497. https://www.ncbi.nlm.nih.gov/pmc/articles/PMC2147080/.

Rosenthal, M.Sara. 2004. *The Skinny on Fat*. Toronto: Macmillan Canada.

Sampson, Wallace. 1995. Antiscience trends in the rise of the "alternative medicine movement". *Annals of the New York Academy of Sciences* 775: 188–197.

Santoni, Matteo, Mimma Rizzo, Luciano Burrattini, Valerio Farfariello, Rossana Beradi, Giorgio Santoni, Giacomo Carteni, and Stafano Cascino. 2012. Present and future of tyrosine kinase inhibitors in renal cell carcinoma: analysis of hematologic toxicity. *Recent Patents on Antiinfective Drug Discovery* Aug 7: 104–10.

Schwartz, R.N., L. Stover, and J.P. Dutcher. 2002. Managing toxicities of high-dose interleukin-2. *Oncology* 16(Suppl 13): 11–20. http://www.cancernetwork.com/oncology-journal/managing-toxicities-high-dose-interleukin-2.

Seuss, Dr. 1960. *Green Eggs and Ham*. New York: Random House Books for Young Readers.

Shehar, Guy. 2012. Interview with Joel Rubin. Cortland Review. http://www.cortlandreview.com/features/12/summer/bruce_joel_rubin_interview.php. Accessed 19 July 2017.

Singleton, J., One9, and Parker, E. 2017. *L.A. Burning: The Riots 25 Years Later*. Entertainment One and A&E Television Networks released April 17.

Stone, Mee-Lai. 2015. The death of Detroit: How Motor City crumbled in the '90s—in pictures. *The Guardian*, February 11. https://www.theguardian.com/artanddesign/gallery/2015/feb/11/the-death-of-detroit-how-motor-city-crumbled-in-the-90s-in-pictures.

Tran, My-Thuan. 2010. Revisiting Haing Ngor's murder: 'Killing Fields' theory won't die. *Los Angeles Times*, January 21. http://articles.latimes.com/2010/jan/21/local/la-me-ngor-murder21-2010jan21.

Turan, Kenneth. 1993. Looking for the meaning of 'Life'. *Los Angeles Times*, November 12. http://articles.latimes.com/1993-11-12/entertainment/ca-55854_1_real-life.

Turhal, Nazim S. 2002. Two cases of advanced renal cell cancer with prolonged survival of 8 and 12 years. *Jpn J Clin Oncol* 32: 152–153.

Walid, M.S., and K.W. Johnston. 2009. Successful treatment of a brain-metastasized renal cell carcinoma. *German Medical Science* 7: Doc28. https://www.ncbi.nlm.nih.gov/pmc/articles/PMC2775194/.

Whorton, James C. 2002. *Nature Cures—The History of Alternative Medicine in America*. New York: Oxford University Press.

Wilkinson, Tracy, and Frank Clifford. 1991. Korean Grocer Who Killed Black Teen Gets Probation. *Los Angeles Times*, November 16. http://articles.latimes.com/1991-11-16/news/mn-1402_1_straight-probation.

World Health Organization. 2017. Definition of Palliative Care. http://www.who.int/cancer/palliative/definition/en/. Accessed 19 July 2017.

"Bye, Bye Life": *All That Jazz* (1979)

4

Film critics and scholars have dubbed *All That Jazz* the "death and dance" film. What is surprising about this end of life film is that it has escaped the attention of most clinical ethicists, and is generally never listed or mentioned as a "bioethics" film per se. This is probably because of its genre: it is a musical. Most notably, the final scene is the riveting dance number "Bye, Bye Life," in which the lyrics are rewritten to the Everly Brothers' original tune, "Bye Bye Love". The new twist on the song replaces "love" with "life" while the familiar "I think I'm gonna cry" is replaced with "I think I'm gonna die."

This is not a typical "bioethics film" in which end of life themes are borne out through drama. This is no ordinary musical, either. Tom Rothman, Chairman and CEO of Fox Filmed Entertainment, states that *All That Jazz* is "unusual in its time; an entirely original musical—written for the screen. It's about death. [Its filmmaker] Bob Fosse's death." Additionally, the musical numbers in this film serve as commentary on the characters rather than as a traditional vehicle for advancing the story (20th Century Fox 2010a).

The Story Behind *All That Jazz*

To understand *All That Jazz*, a brief biography of its filmmaker, Bob Fosse, is in order since the film is an autobiographical account of his own death by heart disease. Aptly put by Rothman: "[*All That Jazz*] is the most conspicuous spectacle ever, of a man stripping himself bare, and of course glorifying himself at the same time, in a movie he is making about himself. Talk about balls…It is the ultimate film about art imitating life imitating art imitating life" (20th Century Fox 2010a). Those who knew Fosse maintain that *All That Jazz* is a too-thinly veiled autobiography.

© Springer International Publishing AG, part of Springer Nature 2018   69
M. S. Rosenthal, *Clinical Ethics on Film*,
https://doi.org/10.1007/978-3-319-90374-3_4

By the mid-1970s, Bob Fosse was at the height of his career. He had won an academy award for *Cabaret* (1972), several Tony awards, with his first for "Pajama Game" (1954), as well as an Emmy in 1973 for the television special *Liza with a Z*. Amidst rehearsing the Broadway musical, "Chicago" and editing a film he was making about comedian, Lenny Bruce titled *Lenny* (1975), the chain-smoking workaholic Bob Fosse had a massive heart attack, and required bypass surgery, which was followed by another heart attack. Fosse thus decided to make a musical about that experience—his near fatal heart attack, his own mortality, and his clear existential angst over his moral lapses throughout his life.

Fosse's personal life was chaotic and self-indulgent. He was a drug user whose infidelity was infamous. Nonetheless, the women in his life remained companions and friends, and two of them even star in *All That Jazz*—Ann Reinking and Jessica Lange. Fosse had cheated on his third wife, Gwen Verdon (who was starring in "Chicago") with dancer Ann Reinking. He then proceeded to cheat on Reinking with actress Jessica Lange, who ultimately left Fosse for the ballet dancer, Mikhail Baryshnikov. Fosse was apparently more upset that Lange left him for a "better dancer" than another man. Gwen Verdon was originally contemplated for the role of the ex-wife in *All That Jazz*. (Verdon was his estranged wife, but they did not divorce.) Ultimately, Leland Palmer played the role of the ex-wife, who emulates qualities of Verdon (20th Century Fox 2010a).

Fosse's first heart attack was a moral inconvenience for all those involved in his life; this is perhaps one of the most critical aspects of this end of life film—unplanned illness and untimely mortality threats can be messy—particularly when there are so many loose threads in one's personal life and social networks. Although Fosse continuously suffered from heart disease and chest pains while making *All That Jazz*, Fosse died seven years later, in 1987, while walking with Gwen Verdon; two of his musicals were being revived on Broadway at the time ("Sweet Charity" and "Cabaret"). He reportedly died in her arms. When asked about his notorious infidelity, Verdon apparently said that Fosse's "real affair was with death" (20th Century Fox 2010a). Fosse cast Jessica Lange as the "Angel of Death" temptress, in fact, and throughout the film, he flirts with Death (Lange) until he succumbs to her temptations.

Ultimately, very few details in *All That Jazz* are fictional. The shots of open-heart surgery that appear in the film are also authentic; four cardiac patients agreed to have their surgeries filmed for the movie (20th Century Fox 2010b). Fosse did wind up changing some names: he calls his surrogate character "Joe Gideon" and called the film *Lenny* "The Standup". Aside from that, most of the details reflect his reality. Fosse was indeed directing a musical ("Chicago") starring his estranged third wife, Gwen Verdon; Joe Gideon is directing a musical starring his ex-wife, played by Leland Palmer. Fosse was having an affair with Ann Reinking when his heart attack occurred; so is Joe Gideon. Fosse is tempted by Jessica Lange (who left him for another man); Gideon is tempted by Lange, who is cast in the role of the temptress/Angel of Death. Fosse's early childhood memories are shown in the film as Gideon's early childhood memories. Fosse has the same heart pains while making the film as Gideon has in the film. Fosse also cast the film with real life characters such as the real film editor for *Lenny*; and the actual lighting designer and stage managers he worked with (20th Century Fox 2010b).

Vincent Canby, in his December 20, 1979 review of the film's opening in the *New York Times* had this to say: "The film is an uproarious display of brilliance, nerve, dance, maudlin confessions, inside jokes, and especially ego. It's a little bit as if Mr. Fosse had invited us to attend a funeral. The wildest show business send-off a fellow had designed for himself and then appeared at the door to count the house. After all, funerals are only wasted on the dead."

Tom Rothman notes "Columbia pictures almost didn't survive this 'dance and death' movie" (20th Century Fox 2010b). Upon Fosse's death, Vincent Canby reportedly said: "History recounts no last words, but it didn't need to; Fosse has already done that himself."

Synopsis

The film centers on the main character, Joe Gideon (Roy Scheider), who is a surrogate for its filmmaker, Bob Fosse. Gideon is a successful choreographer and theatre director who is a chain-smoking workaholic, also addicted to amphetamines. He has a complicated personal life: an ex-wife (Leland Palmer) who is the star of his Broadway musical; a daughter who idolizes him; and a loyal girlfriend (Ann Reinking) he cheats on with various dancers from his show. While working on two intense projects simultaneously—choreographing and casting a new Broadway musical, and editing a film that is vastly over budget and over deadline (called the "The Standup"), Gideon suffers from inconvenient chest pains and is ordered to be on bed rest for a few weeks, jeopardizing the Broadway show. He is not compliant with doctor's orders, and eventually suffers from a massive heart attack and needs to undergo bypass surgery.

The musical numbers are divided between numbers that represent what Gideon is actually doing and choreographing for work (e.g. the opening number "On Broadway" and later, "Take Off With Us") and numbers that are representing his inner reflections on his own mortality and life of moral misdeeds. Self-narrative and reflection dots the film as he confesses his sins and sensational acts to what we presume is an Angel of Death (an attractive temptress, played by Jessica Lange). The film ends with his dying from surgical complications, as the flat lining beeps are intermingled with an outstanding last musical number, "Bye Bye Life"—a musical "eulogy" of sorts.

The Social Location of *All That Jazz*

Fosse's autobiographical film speaks volumes about its timeframe—the 1970s. Although the film was released in 1979, its development and production are actually reflecting the mid-1970s. The decade of the 1970s was a reactionary period to the previous decade. This post-social upheaval timeframe is still processing the social consequences of civil rights, political assassinations, space travel, and the counter-culture/hippie movement. Fosse was born in 1927, and was thus not a baby boomer, but of the same generation of psychologist Timothy Leary (1920–1996), whose "turn on, tune in and drop out" adage had great influence on the young boomers who were in their teens and early 20s during the 1960s. Although the hippie movement is associated with the baby boomers, many older adults, such as Leary, in their 30s and 40s during this time-frame, benefited from, and even exploited, the "Summer of Love" excesses that defined the late 1960s. These older adults were a product of the Beat Generation (Watson 1995), (a.k.a. Beatniks), which flourished in the 1950s, and whose literature is defined by works such as *On the Road* by Jack Karouac (1922–69). (Kerouac was an alcoholic who died from cirrhosis of the liver.)

Yet what we would label as "addiction" today was looked at by Fosse and his contemporaries within the artistic community of the Beat Generation, as romantic, hedonistic indulgence that was part of the creative process. The sense of immortality and indestructibility in the face of tremendous self-harm from alcohol and substance abuse was characteristic of the mindset of this generation (Watson 1995). The Beat Generation was an earlier adopter of drug and sexual experimentation. A large number of individuals in this movement migrated from New York City (the Beatnik epicenter was Greenwich Village) to San Francisco in the late 1950s and became integral to the counter culture movement whose epicenter was the Haight-Asbury district of San Francisco, spawning the "Summer of Love" and the Human "Be-In" in 1967. (The origin of the word "hippie" derives from "hipster" which was first coined by Harry Gibson in 1940 in a song titled "Harry the Hipster".)

Without the predecessor movement of the Beat Generation in the 1950s (Fosse was between the ages of 23–32 during this period), the hippie movement would not have succeeded (Watson 1995; Miller 1991). Thus, the hippie movement's patron saint is the Beat Generation—Fosse's generation—born between 1920 and 1939. Others in this generation include Diane Arbus (1923–71); Lenny Bruce (1922–66), also the subject of one of Fosse's films; Truman Capote (1924–84); and Judy Garland (1922–69)—all addiction-related deaths (Collins and Skover 2002; Lublow 2003; Clarke 2001). Other notable hedonistic figures in this generation include Andy Warhol (1928–87); Ken Kesey (1935–2001), a major advocate of LSD, and author of <u>One Flew over The Cuckoo's Nest</u> in 1962 (made into a film in 1975, discussed in Chap. 6), and, unfortunately, Charles Manson (1935–2017), who infamously used LSD and hedonism to form a large cult of sex slaves who ultimately murdered for him.

Understanding Fosse and *All That Jazz* within the social context of the *aging* hedonist from the Beat Generation informs us about Fosse's love affair with self-indulgence and little regard for its consequences. His chain-smoking, pill popping, and womanizing are condoned by his peers. Ultimately, Fosse's heart attack in real life was his mortality moment; *All That Jazz* lays bare his moral reflections regarding his physical and emotional indulgences.

One of the film's central messages is that abusing one's body can lead to untimely and inconvenient death. This message is a new one for its 1979 audience still basking in the glory of newly found civil and (women's) reproductive rights—the Pill, access to abortion, and gay liberation. Politically, the United States in the aftermath of Vietnam and Watergate had become disillusioned with its government, institutions and military heroes, which gave rise to blatant disregard for rules and authority. The disastrous Carter administration (1976–80) is largely viewed as impotent and irrelevant, creating a devaluing of American leadership and ideals. Other films released during 1979 echo this disillusionment: *Apocalypse Now, And Justice for All, The China Syndrome, Norma Rae, Being There,* and *The Great Santini.* Even *Alien,* which is a "haunted house in space" horror film resonates this disillusionment, as the plot focuses on a strong heroine who has been betrayed by the corporation in charge of her safety and spacecraft.

In 1979, most young adults in their 20s and 30s are having unprotected sex with multiple partners, and there is no clear definition of sexual abuse, date rape, and certainly no "#Me Too" movement; in the gay community, promiscuity is celebrated (Tyrkus 1997). In 1979 "safe sex" means not having sex in a moving vehicle; no one is fearful of AIDS (the first case of AIDS is not reported until 1981). On the drug front, young adults are moving from hallucinogenic drugs to mood enhancers such as cocaine (a drug for the rich), and marijuana is a social norm (NDADD 2017). On the controlled substances front, everybody smokes and drinks, while "smoke-free" public spaces are unheard of (National Commission 2017). Studio 54 is still in its heyday in 1979, which is essentially a modern "opium den" for its age.

Meanwhile, normative social institutions are falling apart, and adults are beginning to delay marriage and childbirth (the U.S. birth rate dramatically drops in the 1970s); they question the values associated with family and marriage. Other films released in 1979 echoing the de- and re-construction of traditional social institutions include *Kramer v. Kramer, Manhattan,* and *10.*

Named the "Me Decade" by Tom Wolfe in a 1976 issue of *New York Magazine* (Wolfe 1976), the 1970s opens with many untimely deaths related to hedonistic behaviors. The deaths of Janis Joplin (1943–70), Jimmy Hendricks (1942–70) and Jim Morrison (1943–71) are all representative of hedonistic drug-related deaths. There are two films released in 1979 that begin to re-examine a misguided romanticism associated with hedonism, and its self-destructive underbelly of addiction: *The Rose,* which is based on the life of Janis Joplin, and *All That Jazz.* The topic of hedonistic behavior clearly fascinates Fosse. He begins to examine this behavior in his film, *Lenny* (1974), about one of his contemporaries who dies from addiction (Collins and Skover 2002). But to truly understand and appreciate the context of addiction in the 1970s, I suggest that there is a distinction between the consequences of *hedonistic drug use,* which these films explore, and self-medicating with drugs to palliate painful social location (e.g. poverty and racism), illustrated in films such as *Lady Sings the Blues* (1972), a biography of Billie Holiday.

Ultimately, addiction films in the 1970s, such as *Panic in Needle Park* (1971) are few and far between, as cocaine is still in its ascent before it's repackaged as crack in the 1980s; a 1979 *New England Journal of Medicine* paper makes the first clinical warning about the new trend of "smoking cocaine" (Siegel 1979). Meanwhile, a 1979 *Journal of the American Medical Association* paper concludes that "recreational" cocaine use is not safe (Welti and Wright 1979). On the health front, the baby boomers are feeling invincible. But even more ahead of its time *All That Jazz* tackles what is certainly not discussed in the 1970s: health behaviors that lead to cardiovascular disease.

The History of Medicine Context

A film about heart disease related to bad personal health habits is ahead of its time in 1979. Although it was recognized by the 1940s that heart disease is the number one killer in the U.S., its lifestyle and dietary triggers are not well understood until the 1980s (Pampel and Pauley 2004).

One pioneer, Ancel Keys, MD, predicted the "diabesity/heart disease" epidemic as early as 1959; his book, Eat Well and Stay Well proposed dietary and lifestyle guidelines for a "healthy heart" that were dismissed by his peers as radical and unsubstantiated; ironically, Keys' guidelines are identical to today's dietary guidelines: maintaining normal body weight; restricting saturated fats and red meat; using polyunsaturated fats instead to a maximum of 30% of daily calories; plenty of fresh fruits and vegetables and non-fat milk products; avoidance of overly salted foods and refined sugar; exercise, lower stress, and quit smoking. Unfortunately, no one was listening to this medical outlier; Keys would not get credit for his work until the 1990s (Rosenthal 2004).

In 1948, a thirty-year heart study began in Framingham, Massachusetts, which was known as the Framingham Heart Study (NIH 2017). This was the first longitudinal study conducted in the U.S. on heart disease and enrolled 5127 people aged 30–62 who showed no signs of heart disease, who were examined every two years and followed for signs of heart disease so triggers could be better understood. By 1961, five years' worth of data revealed men under 50 with elevated cholesterol, high blood pressure, and high blood sugar were at greater risk of heart disease, but the lifestyle habits contributing to these precursor conditions were not well understood. Lifestyle factors such as diet, smoking, and lack of exercise would not be understood as clear triggers of heart disease in middle-aged men until much later; obesity was not even clearly established as a risk at this point. In the medical research community, what was known as the Framingham "risk factors" for heart disease led to further research in looking for clues to lifestyle triggers. By 1967, the World Health Organization stated that heart disease was the world's most serious epidemic (Pampel and Pauley 2004). In 1974, 24 years of data from the Framingham study revealed more data surrounding cholesterol levels and men, leading to more research surrounding the link between diet and heart disease, but concrete evidence linking smoking as a cause of heart disease was not yet published. Chain-smoking Fosse read on his cigarette packages (Dumas 1992): "Caution: Cigarette Smoking May be Hazardous to Your Health" (1966–1970) and "Warning: The Surgeon General Has Determined that Cigarette Smoking is Dangerous to Your Health" (1970–1985). It was not until 1985 that warning labels on cigarette packages would specifically list the causal link of smoking to heart disease (Dumas 1992; CDC 2018; National Commission 2017). Primary prevention of heart disease would not become a focus in mainstream medicine until the 1980s. Smoking cessation counseling was generally not a routine topic of discussion between doctors and their smoking patients, either.

Treatment of cardiovascular disease in 1979 relies on surgical interventions, which begin to flourish in the 1970s. The first heart bypass surgery is performed in 1960 at the Cleveland Clinic, and begins to be offered as a standard of care in the 1970s; the first heart transplant is done in South Africa in 1967, and then three days later, in the U.S., by the same surgeon who invented the left ventricular assist device (LVAD) in 1972, and then the intra-aortic balloon pump that would not be in general use until the 1980s (Hoffman 2008). What is not generally dealt with, however, is surgical candidacy for these procedures, as some candidates would have better outcomes than others. Today, smokers must demonstrate that they have stopped smoking before they undergo such procedures, which was not the case in 1979.

It would not be until 1981 when Bruce Reitz of Stanford University in California would perform the first successful heart-lung transplant (Reitz 2011). The key to his success, however, was his experimenting with the drug, Cyclosporine to deal with the problem of rejection. Thus transplant surgery was truly not a viable option for cardiovascular patients until the late 1980s until rejection medication was perfected.

From a public health standpoint, the root causes of cardiovascular disease are not being dealt with at all in the 1970s. For example, even though there was some scientific evidence that linked smoking and heart disease in 1967, there was still not enough evidence to state that it was an absolute cause of heart disease or warn the public. Only in lung cancer and emphysema were there established causal links. Even so, there were no social restrictions on smoking whatsoever. Smokers abounded in workplaces, and all public places—including many hospitals. The only exception to this was the state of Minnesota, which enacted the *Minnesota Clean Indoor Air* Act in 1975, making it the first state to ban smoking in most public spaces; however, Minnesota restaurants were required to have smoking sections and were not smoke free until 2007 (National Commission 2017). Most smoking bans would not be enacted in U.S. states until after 2000, with only a trickling of clean air legislation in the 1990s. Smoking bans on airlines do not begin until the late 1980s, and only in 2000 are all flights made smoke free by mandated legislation (Americans for Non-Smokers' Rights 2017).

One of the first congressional bans on tobacco advertising on television passes in 1971; however, marketing experts within the tobacco industry survey the youth of the period (their target audience), and conclude that the attitudes of youth in 1971 is much more hedonistic (what they cite as a "live dangerously" philosophy) and companies are not worried about losing this key demographic despite mandated warning labels and bans on television advertising.

The vast majority of Americans are unaware of the causes of heart disease in 1979; this would not become a well-covered health topic for mainstream media until the mid-late 1980s and 1990s, when popular heart health gurus such as Dr. Dean Ornish begin to penetrate into the public consciousness. In 1976, and 1978, respectively, two dietary "atom bombs" are covertly delivered into the American diet: hydrogenated soybean oil (1976) and high fructose corn syrup (1978). These ingredients become common in most packaged foods by 1985, which help to create soaring obesity and diabetes levels. However, tobacco's clear threat to the human heart remains at the sidelines as a public health message into well into the 1990s (Dumas 1992).

Even less understood until the 1990s was the effect of recreational drug use and addiction on the heart. Addiction research is in its infancy, with a focus on alcoholism. In the mid-1970s, there is debate as to whether treatment for drug abuse and alcoholism should be regarded as one discipline, and clinically and administratively merged (NCADD 2017; White 1998).

In the film, Joe Gideon begins each morning with dextroamphetamine and sodium bicarbonate, chain smokes, and never seems to sleep. Excessive drinking and other recreational drugs dot his social activities. Without the luxury of his medical records, one can speculate that Bob Fosse's first heart attack in his 40s, and ultimate death from his third heart attack at age 60, was probably triggered by drug and tobacco abuse. In fact, drug information labels for dextroamphetamine post serious warnings that "abuse of dextroamphetamime may cause serious heart problems, blood vessel problems, or sudden death." Of course, these warnings would not be widely available to 1970s patients, who are not routinely accessing an internet that didn't yet exist, or even going to medical libraries to look up drug information in the Physician's Desk Reference.

Given the state of knowledge regarding cardiovascular disease risk factors, *All That Jazz* sends a message to its audience about predictive health that stands alone in its time. In its long "end of life" musical sequences toward the end of the film, a "Whose Sorry Now?" number makes it clear that Fosse is well aware of the causal link between his bad health behaviors and his heart; it refers to too much "boozing, [drugs], smoking, and sex". We must be mindful that Fosse is telling his original audience something that would not become mainstream advice until at least a decade later.

Clinical Ethics Themes

Healthcare providers who deal with end of life have all had Joe Gideon as a patient. Fosse actually provides the healthcare provider perspective in two particular scenes: one which appears to be a case conference in which the healthcare team wrestles with how to handle Gideon as a patient; another in which Gideon's frustrated cardiologist confronts Gideon about his non-compliance: "If *you* don't give a damn, it's hard for *us* to." Joe Gideon is a very familiar patient to cardiologists, who typically deal with self-destructive patients since the vast majority of heart disease is linked to poor health behaviors.

Joe Gideon is thus the "difficult patient" who is self-destructive and non-compliant. Such patients are willfully exercising their right to self-determination by making extremely bad judgments when it comes to their health. At the same time, these are patients whose poor health behaviors are enabled by their societies and environments (Browne et al. 2003). As I point out to my medical students, can we genuinely berate patients who smoke when tobacco is a legal, and a promoted substance in our society? And then, once they are addicted (the result of aggressive marketing), do we treat the addiction as a disease or a behavior? In a more current venue, can we berate obese patients when the food industry has conspired to aggressively market and sell us fattening and unhealthy foods, or make such foods more affordable than healthful foods (hence, the diets of poverty)? The rhetoric of personal responsibility is difficult to preach when our government institutions protect the tobacco and food industries. Similarly, patients who overuse prescription medications may be enabled by unscrupulous physicians. Patients, on the other hand, who use illegal drugs, are more justifiably self-destructive. Joe Gideon demonstrates the common problem, too, of polydrug use, in which he is dependent on multiple substances that may interact with one another. For a current audience, Gideon's behaviors echo a too-familiar problem of cardiomyopathy directly consequent to methamphetamine and crack/cocaine use. Such patients will typically require repeated heart valve repair.

When viewing this film from the practitioner lens, the clinical ethics dilemma highlighted is the issue of beneficent care for this self-destructive cardiology patient. Beneficence refers to maximizing clinical benefits and minimizing clinical harms. A non-compliant, chain-smoking patient, for example, may not be a suitable candidate for heart bypass surgery. The question I would ask when viewing *All That Jazz* today is: "why is he even being considered for bypass surgery?" While this is a patient who might actually live longer if only he would listen to the advice of his practitioners, he is also a patient who may benefit from a completely different approach—a palliative care approach. When patients' behaviors put them at greater risk during surgery, it cannot necessarily be said that a life-saving surgery truly "maximizes" benefits when risks of dying on the table may increase to well over 50%.

In cases such as the self-destructive, younger patient whose health behaviors are threatening to end their lives prematurely, there is frequently a failure amongst healthcare providers to craft appropriate end of life discussions and address end of life issues for patients. Instead, the youth of the patient, and his or her "death defying" behaviors, inspire healthcare providers to offer overly aggressive thera-pies. Such patients are frequently very poor surgical candidates, for example, and may wind up requiring repeated surgeries, for little benefit. Beneficent care may not be achieved through surgery or aggressive care. Would Joe Gideon have benefited from a more palliative care approach, in which more effective closure with friends and family could have been facilitated? It seems he would have benefited more from a "living wake" with his close companions around him. In the film, he dies on the table undergoing risky heart surgery. The entire film is framed in the context of the patient's imaginary closure while awaiting a risky surgery in which death is a real possibility. As a clinical ethicist, I have personally been involved in cases where I ask the cardiology team and cardiac surgeons to think through all options, and not to dismiss the options of palliative care and social supports for self-destructive patients in whom surgery would be of little benefit but who still have existential suffering (Rousseau 2001).

In fact, Fosse is clearly thinking the same thing.

The Celebrity Patient

A novel theme presented in *All That Jazz* is the vulnerability of the "celebrity" or "VIP" patient, who perhaps should be treated as a distinct vulnerable population. In a comic scene that touches on professional ethics, Gideon's "Broadway Doctor" (who is hired by the producers to ensure a clean bill of health to the financial backers of the show) does an incompetent exam on a coughing and hacking chain-smoking Gideon, while chain-smoking and coughing himself. (In fact, I show this scene to medical residents when I teach ethics and professionalism, to illustrate poor clinical conduct while examining a patient.) The "Broadway Doctor' clearly ignores his ethical duties and proclaims Gideon to be in "perfect health". This scene remains just as relevant today in the wake of egregious ethical lapses by physicians who irresponsibly prescribe medications to addicted celebrities; falsely certify or attest to their good health for financial gain, or fail to appropriately refer celebrity patients because they want to "claim" them as their own. Such themes are present in many celebrity deaths, ranging from Marilyn Monroe and Anna Nicole Smith to Michael Jackson, whose cardiologist (not board certified) provided the dangerous drug, propofol, to Jackson, which led to Jackson's death. Ultimately, how do we treat talented, narcissistic patients who are famous?

Existential Suffering and the Patient's Perspective

Existential suffering at the end of life refers to the "moral residue" brought on by an absence of closure with friends or family, and regrets of past behaviors or incomplete life journeys. If nothing else, *All That Jazz* invites the audience into Fosse's moral residue. The film is literally Fosse's catharsis, in which he relieves his existential suffering. Fosse seeks absolution on screen for what he has not sought out in real life. The participants in the production of *All That Jazz* reported that despite the film's intensely personal content, Fosse remained an "asshole"— egotistical, unapologetic, and very difficult to work with (20th Century Fox 2010b).

Richard Dreyfuss, who was cast early as the Joe Gideon character after an exhaustive search for a lead, quit the production because he found Fosse so impossible to work with. (Dreyfuss had his own battle with addiction, discussed in Chap. 2.) Fosse was so egocentric that he even lobbied to cast himself in the role after Dreyfuss quit, which was prevented by the producers, as it surely would have led to disastrous results. Roy Scheider, the "other *Jaws* star" was ultimately cast, and is perfection in the role (20th Century Fox 2010b).

Existential suffering is an area of end of life care that did not become timely until the 1990s (Byok 1996), when it demanded more attention. Pastoral and divinity experts are typically called upon to discuss moral burdens with dying patients. One could argue that the sole purpose of *All That Jazz* was a "make work" project Fosse created to deal with his existential suffering after his "mortality moment" occurred when he had his heart attack.

All That Jazz is unique in the end of life genre because it is more about the threat of "sudden death"—and the moral inconvenience of unfinished business—than it is about a patient slowly coming to terms with a terminal illness (as seen in *Wit* or *My Life*, for example). Joe Gideon's death, in fact, leads to a train wreck for surviving family and colleagues, who are seriously inconvenienced by his passing. This is one of the reasons for so much of Fosse's/Gideon's existential suffering. He knows he is not leaving his life tied up in a neat package, but leaving when things are considerably messy.

The last shot of the film perhaps does tidy things up: We see Gideon's body being neatly zipped up into a corpse bag in the morgue. The credits then roll to a jerking, almost inappropriate, Ethel Merman version of "There's No Business like Show Business". Fosse clearly wants to die onstage, and makes it so. In reality, he dies in 1987 while walking down Broadway with his ex-wife, Glen Verdon.

Film Stats and Trivia

- Ranked 14 out of AFI's best 100 movie musicals.
- 9 academy award nominations.
- Bob Fosse is the first person to have won an Oscar, Tony and Emmy for best director.
- Co-Writer of the film, Robert Alan Aurthur died during the making of the film.
- Richard Dreyfuss was originally cast but dropped out because of the difficulty he had working with Fosse.
- Fox joined with Columbia to save the finances of *All That Jazz*, making the film the first to share studio credits.
- *All That Jazz* was not a great financial success, earning 1.5 million for its investors.

From Theatrical Poster

Directed by Bob Fosse
Produced by Robert Alan Aurthur
Written by Robert Alan Aurthur and Bob Fosse
Starring: Roy Scheider; Jessica Lange; Leland Palmer; Ann Reinking
Music by Ralph Burns Cinematography Giuseppe Rotunno Editing by Alan Heim
Distributed by 20th Century Fox
Columbia Pictures Release date(s) United States:
December 20, 1979
Also starring:
Cliff Gorman as Davis Newman
Ben Vereen as O'Connor Flood
Michael Tolan as Dr. Ballinger
John Lithgow as Lucas Sergeant

References

20th Century Fox. 2010a. All That Jazz, Part 1. *Fox Legacy with Tom Rothman*. Season 1, Episode 42.

20th Century Fox. 2010b. All That Jazz, Part 2. *Fox Legacy with Tom Rothman*. "Season 1, Episode 43.

Americans for Non-Smokers' Rights. 2017. Smokefree Transportation Chronology. http://nosmoke.org/document.php?id=334. Accessed 22 Feb 2018.

Browne, A.B., B. Dickson, and R. Van Der Wal. 2003. The ethical management of the noncompliant patient. *Cambridge Quarterly of Healthcare Ethics* 12: 289–299. https://www.cambridge.org/core/journals/cambridge-quarterly-of-healthcare-ethics/article/ethical-management-of-the-noncompliant-patient/D902732A57B91BB4CC86A2905679860.

Byok, Ira. 1996. The nature of suffering and the nature of opportunity at the end of life. *Clinics in Geriatric Medicine* 12: 237–252.

Canby, Victor. 1979. The Screen: Roy Scheider Stars in 'All That Jazz': Peter Pan syndrome. *New York Times*, December 20.

Centers for Disease Control. (CDC). 2018. Warning labels. https://www.cdc.gov/tobacco/data_statistics/sgr/2000/highlights/labels/index.htm. Accessed 24 Feb 2018.

Clarke, Gerald. 2001. *Get Happy: The Life of Judy Garland*. New York: Random House.

Collins, Ronald K.L., and David M. Skover. 2002. *The Trials of Lenny Bruce: The Fall and Rise of an American Icon*. Naperville, IL: Sourcebooks Mediafusion.

Dumas, Bethany K. 1992. Adequacy of cigarette package warnings: An analysis of the adequacy of federally mandated cigarette package warnings. *Tennessee Law Review* 59: 261–304.

Framingham Heart Study. 2017. A Project of the National Heart Lung and Blood Institute and Boston University. www.framinghamheartstudy.org/. Accessed 22 Feb 2018.

Hoffman, Jascha. 2008. Dr. Adrian Kantrowitz, cardiac pioneer, dies at 90. *New York Times*, November 19.

Keys, Ancel, and Margaret Keys. 1959. *Eat Well & Stay Well*. New York: Doubleday.

Lubow, Arthur. 2003. Arbus reconsidered. *New York Times Magazine*, September 14. http://www.nytimes.com/2003/09/14/magazine/arbus-reconsidered.html?pagewanted=9.

Miller, Timothy. 1991. *The Hippies and American Values*. Knoxville: University of Tennessee Press.

National Commission on Marihuana and Drug Abuse. 2017. History of Tobacco Regulation. http://www.druglibrary.org/schaffer/library/studies/nc/nc2b_10.htm. Accessed 22 Feb 2018.

Pampel, Fred C., and Seth Pauley. 2004. *Progress Against Heart Disease*. Westport, Connecticut: Praeger.

Reitz, B.A. 2011. The first successful combined heart-lung transplantation. *The Journal of Thoracic and Cardiovascular Surgery* 141:867–869. http://www.jtcvsonline.org/article/S0022-5223(10)01445-5/abstract.

Rosenthal, M.Sara. 2004. *The Skinny on Fat*. Toronto: Macmillan Canada.

Rousseau, Paul. 2001. Existential suffering and palliative sedation: A brief commentary with a proposal for clinical guidelines. *American Journal of Palliative Care* 18: 151–153.

Siegel, R.K. 1979, Cocaine smoking. *New England Journal of Medicine* 300:373. http://www.nejm.org/doi/full/10.1056/NEJM197902153000730.

The National Council on Alcoholism and Drug Dependence. 2017. Significant Events in the History of Addiction Treatment and Recovery in America. https://www.ncadd.org/about-ncadd/about-us/timeline-of-events. Accessed 22 Feb 2018.

Tyrkus, Michael J. 1997. *Gay & Lesbian Biography*. New York: St. James Press.

Watson, Steven. 1995. *The Birth of the Beat Generation: Visionaries, Rebels, and Hipsters, 1944–1960*. New York: Pantheon Books.

White, William. 1998. *Slaying the Dragon: The History of Addiction Treatment and Recovery in America*. Bloomington, IL: Chestnut Health Systems.

Welti, Charles V., and Robert K. Wright. 1979. Death caused by recreational cocaine use. *Journal of the American Medical Association* 241: 2519–2522.

Wolfe, Tom. 1976. The "Me" decade and the third great awakening. *New York Magazine*, August 23. http://nymag.com/news/features/45938/.

Part II
Films About Competency and Decision-Making Capacity

It is said that the "bread and butter" of a clinical ethics consultation service revolves around decision-making capacity questions and concerns. Unfortunately, the terms "capacity" and "competency" are often conflated in medical charts, and among healthcare providers, *but they are two different things*. As discussed in detail in the chapters of this section, *competency* is a legal term that refers to any emancipated adult who has not otherwise been assigned a guardian by the courts. In these individuals (most of us), autonomy, and thus, decision-making capacity is presumed until proven otherwise. Competency can be viewed as a characteristic of a person that does not change over short periods of time. Children, for example, are not legally competent until they are 18. A severely developmentally delayed adult with congenital brain damage and special needs will never be competent, and will always require a guardian. A patient with severe mental health problems that prevent rational decisions may also be legally incompetent; a person with severe dementia may be declared legally incompetent. A person in a persistent vegetative state may be declared legally incompetent. But some legally incompetent patients may be able to make some medical decisions, and thus, may have varying degrees of decision-making capacity based on the U-ARE criteria (see Table 5.1), which stands for Understanding, Appreciation, Rationality and Expression of a choice or preference.

Although decision-making capacity is an ability that competent adults have, it may change over short periods of time due to many barriers, which can be caused by physiological changes (severe pain, metabolic changes, etc., bodily trauma), altered mental status (mood disorders, acute psychiatric illnesses, substance use/abuse, neurological changes, etc.), or psychosocial barriers (language and literacy). In a healthcare context, patients must be able to make *medical decisions,* but they don't necessarily need to know what day it is, if they have been lingering in an ICU for weeks, but have very clear preferences, such as "no feeding tubes".

It gets grayer. Some patients declared "incompetent" can still make some of their medical decisions, such as a 17 year-old, or a person with dementia who has periods of lucidity. For many of these shades of gray, clinical ethicists suggest Assent. Patients agree, in general principle, to the goals of care, but may not be able to consent to every small detail; such patients still need a surrogate decision-maker (or parental authority) to be their voice or to sign legal documents. At the same time, there are many competent patients who lose decision-making capacity for either short periods of time or long periods.

Similarly, there is a lot of confusion over who should be making medical decisions. In the healthcare setting, we look for the obvious documents first: court-appointed guardian or a document that grants then medical Power of Attorney. When there are no such documents, in the United States, we next search for an Advance Directive of some kind that may have named a surrogate. No such luck? We next use the family hierarchy system to select a surrogate decision-maker; sometimes, family members are the worst people to make decisions because of secondary gain. In cases where the patient has no representation, guidance varies by country or state. In many U.S. states, hospital employees are presumed to have a conflict of interest and thus, a state-appointed guardian is required—a system that is overloaded and often flawed.

Each film in this section prompts discussion of competency, decision-making capacity, the U-ARE criteria, and surrogate decision-making. *Diving Bell and the Butterfly*, based on a real case, is about a French patient with locked-in syndrome, who is "betwixt and between", and eventually, enabled to become fully decisional. *One Flew Over the Cuckoo's Nest* is about a prisoner patient looking to "fake" incompetence to get out of his prison sentence, and winds up involuntarily committed at a psychiatric hospital in the process. *Still Alice* is about a woman with early-onset Alzheimer's disease and her slow-motion neurological decline.

Core Bioethics Principles Involved

Ultimately, the core bioethics principle dominating in this section is the *Principle of Respect for Persons,* in which the healthcare provider must protect patients with either limited or unproven autonomy due to their clinical presentations, and try to honor any previous or known preferences, where possible. This involves ensuring that there is an authentic surrogate decision-maker involved, who can make decisions based on either known patient preferences or at least, in their best interests. In *Diving Bell and the Butterfly*, we see a beautiful (and ideal) example of a stroke patient who is fully enabled to communicate his decisions by exhaustive efforts of the healthcare team. In *One Flew Over the Cuckoo's Nest*, we see psychiatric abuse in full bloom when competent and capacitated adults are declared "incompetent" because of their social behaviors. In *Still Alice*, we are struck by the pre-dementia self and the post-dementia self: they are not the same people and do not choose the same things, and it's difficult to see the bright line of when, exactly, this patient loses capacity.

The Principle of Beneficence obligates healthcare providers to maximize clinical benefits and minimize clinical harms by typically ensuring there is a greater balance of benefits than harms. In *Diving Bell and the Butterfly*, the patient objects to his eyelid being sewn shut, but cannot speak; in an emergent situation, the healthcare providers are guided only by Beneficence, and proceed. We still harm people when guided by Beneficence. In *One Flew Over the Cuckoo's Nest*, an extreme behavioral threat to the staff (the patient attempts to choke a nurse to death) is lobotomized—presumably for the "good" of all involved. In *Still Alice*, the family makes decisions in her best interests in the end, and most of us are left troubled by the results. But we can also see ourselves making similar plans.

The Principle of Non-maleficence is the explicit obligation not to knowingly offer a treatment that has no benefit, or to harm patients; again, this is often a conjoined principle with Beneficence as there may not be a bright line where violating Non-maleficence is an issue. Lobotomy used to be viewed as "beneficent", for example.

Finally, the Principle of Justice in clinical ethics looks at healthcare access and resources. When we look at the films in this section, is resource allocation interfering with determining, or enabling capacity? We must ask how American patients with locked-in syndrome would be treated in an average hospital; or, are we still using psychiatric treatments to regulate social behaviors given limited staffing resources to deal with them? Are we asking too much of families when we ask them to take on the burden of full-time caregiving?

One Eye and an Alphabet: *Diving Bell and the Butterfly* (2007)

5

This film is based on the autobiography of stroke patient, Jean-Dominique Bauby, who was the Editor of the French-based fashion magazine, *Elle*. This film is about a successful endeavor by French healthcare providers to find a way to communicate with Bauby, who had "locked in syndrome". He was completely paralyzed, but had the ability to blink his left eyelid. Due to his paralysis, his right eyelid needed to be sewn shut because it was not closing or blinking adequately, and was at risk of drying out.

Bauby had lost his capacity to communicate, yet had a fully lucid mind, and was completely capable of making decisions: he just couldn't express himself, and thus, did not meet the criteria for decision-making capacity at first glance, but for one eye that he could blink with. With the help of dedicated and persevering healthcare providers, Bauby is introduced to a method of blinking to letters in an alphabet to spell words and communicate. His ability to communicate transforms his quality of life. He wrote a short memoir using that method of blinking, entitled Diving Bell and the Butterfly (Bauby 1997), which is an account of how it feels to have "locked in syndrome". When Decision-Making Capacity and Competency is on the menu in any medical ethics curriculum, this is a most illuminating and unique foreign film to view, which reveals that there is more to capacity and competency than, shall we say, "meets the eye". This film is mostly shot from the viewpoint of Bauby's left eye, so the viewer sees what he sees. But the other part of this story is a discussion about the French healthcare system, which was ranked in 2000 by the World Health Organization as the best healthcare system in the world. This story, in some ways, could only happen in the U.S. if the patient were a "billionaire". But in France, the extraordinary care that Bauby receives is available to all French citizens. Thus, *Diving Bell and the Butterfly* is about one stroke patient's experience in a superb healthcare environment, who ultimately died, but had the best outcome one could hope for, given the extreme limitations of his condition. So this film requires a discussion of the disparities in stroke care in France versus the U.S. because it is not an American experience. Finally, at its heart, this film is about enabling communication to activate autonomy.

© Springer International Publishing AG, part of Springer Nature 2018   87
M. S. Rosenthal, *Clinical Ethics on Film*,
https://doi.org/10.1007/978-3-319-90374-3_5

Thus, a review of Speech Language Pathology (SLP) as a field, and other communicative disorders is another important theme, and this film may be the best clinical ethics film of all to select for healthcare trainees in SLP, as it's a case study in SLP ethics as well.

Origins of *Diving Bell and the Butterfly*: The Book

Jean-Dominique Bauby was 43 years-old when he suffered a massive stroke December 8, 1995. After being in a coma for 20 days, and awake but incapacitated for several weeks in the hospital, he slowly began to regain awareness in a completely paralyzed body, including his mouth, arms and legs. Bauby had lost 60 lb during the course of his coma. He was formally diagnosed with the neurological condition of "locked in" syndrome (LIS), where it was clear (based on neurological tests and scans) that his thinking was fully functional, but not his body. Bauby remains in this state until his death in 1997, but remarkably, regains the ability to communicate with skilled speech language pathologists who use an alphabet card that allows him to "blink" to letters forming words. In his previous life he is a busy Editor of the fashion magazine, *Elle*, in which he is at the forefront of Paris fashion in a very fruitful period: the 1980s and 1990s. Bauby, prior to his stroke, had a publishing deal to write a book about a female "count of Monte Crisco" (Bauby 1997). He decides instead to write a memoir about his experience as a patient with locked-in syndrome, which is also very much a "prison" theme nonetheless. Most ironic, Alexander Dumas' famous tale is noted in the neurological community for describing the effects of stroke in two incidental characters in the book. The character of Monsieur Noirtier de Villeforte, is described as having "locked in" syndrome in this way (Williams 2003):

> Sight and hearing were the only two senses which, like two sparks, still lit up this human matter, already three quarters moulded for the tomb. Moreover only one of these two senses could reveal to the outside world the inner life, which animated this statue.... He was a corpse with living eyes, and at times, nothing could be more terrifying than this marble face out of which anger burned or joy shone. (Dumas 1996: 564)

Dumas' character is able to communicate through blinking, closing his eyes once for 'yes' and several times for 'no'. His granddaughter, Valentine, uses a dictionary to help him spell out letter by letter what he wishes to say (Williams 2003). Alexander Dumas' original edition of The Count of Monte Cristo (1844) was described as having what we now understand to be LIS—all the more ironic, considering that Bauby's original book idea was to rewrite that story with all female characters. If this film is selected as part of a Narrative Medicine course (or even a French literature course), it may be nice to include passages or the entire work of The Count of Monte Cristo, as this is a French literary classic work that Bauby connects with and identifies with to the point where he was contracted to write about it before, literally, embodying it in some ways.

Bauby's publisher assigns him a dedicated translator, who will use the alphabet system, which entails reciting individual letters that Bauby would blink to, which would form the words and sentences. This was a very laborious practice, but it led to a book. (This is known as partner-assisted scanning—see under History of Medicine.) To make the alphabet recitation go faster, Bauby's dictation assistant, Claude Mendibil (who is thanked in his memoir), recited the alphabet letters based on their usage frequency. Diving Bell and the Butterfly (in French: Le Scaphandre et le Papillon) is so named because Bauby feels as though he is weighted down by the old fashioned "diving bell" used for divers; but he can free himself with his thoughts, which he compares to the butterfly. States Bauby: "My heels hurt, my head weighs a ton, and something like a giant invisible diving bell holds my whole body prisoner...My diving bell becomes less oppressive, and my mind takes flight like a butterfly. There is so much to do. You can wander off in space or in time, set out for Tierra del Fuego or for king Midas's court." (Bauby 1997: 3, 5). Bauby eventually completes his memoir, but dies of pneumonia two days after its publication.

In 1997, shortly after the book's publication, a documentary about Bauby was released in France, entitled *Assigné à Residence* ("Assigned Residence") about Bauby's hospitalization (Beineix 1997); the documentary featured his writing assistant/dictator, Claude Mendibil, who Bauby thanks in his dedication: "And my deepest gratitude to Claude Mendibil, whose all-important contribution to these pages will become clear as my story unfolds" (Bauby 1997). It also featured Bauby's significant other; his personal life was complicated. The book is quite short, and is not noted as much for its literary intrigue as its process of composition. According to one critic, the prose in the book "are merely competent—a compelling collection of incidents, but little more. 'The Diving Bell and the Butterfly' stands out from the crowd because of the circumstances of its creation, but Bauby, for all his self-expressed disdain for the Elle editorship, still writes in episodic, glossy-ready bits and pieces. Even isolated in the seaside Naval Hospital at Berck-sur-Mer, Bauby's mind is never far removed from the palatability of weekly-mag kitsch..." (Uhlich 2007).

Bauby had two children (son, Theophile, and daughter, Celeste) with his girlfriend, Sylvie de la Rochefoucauld. After Sylvie and Bauby ended their relationship, he started seeing Florence Ben Sadoun, who was his girlfriend at the time of his stroke, and who faithfully visited him at the Berck-sur-mer. Sadoun was featured in the 1997 documentary as well. Bauby also discusses Sadoun in his memoir.

Of note, Bauby's hospital was in a resort setting on the French Channel: the Naval Hospital at Berck-sur-Mer (Bauby was in room 119). I discuss the hospital in more detail further on (see under Social Location).

The Film

The film version was released in 2007, and was initially a Universal Studios project with Johnny Depp set to play Bauby; British screenwriter, Ronald Harwood, wrote the screenplay. Depp suggested Julian Schnabel as director. But the project fell apart because Depp had a conflict with another film. Universal Studios withdrew, and the project was renegotiated with Pathe studios in the U.S. Pathe wanted to do a bilingual project, filming one version in French and one in English, but Schnabel decided to just film a French version (shown with English subtitles in the U.S.), using an all French cast. Schnabel states: "The book was written in French, by a Frenchman, in a French hospital… And I'd be damned if English and American people were going to make believe they are French" (Hartman 2007). He persuaded the producers that the French version was the only culturally competent version to make, and in teaching this film, I agree.

Schnabel was both a painter and director, so made much of the beautiful scenery of the hospital location (see further). Schnabel stated that his interest in the film came from personal experience; he had read the memoir on his own two years before receiving the script. Then, when his 92-year old father was very sick battling prostate cancer, the script arrived. He said (Stock 2008):

> I knew about the book for years, but I had no intention of making this movie. I'd been terrified of death my whole life - like my father. And I guess I was next on the conveyor belt, so I figured I'd better straighten this out before I go… So I made this movie and you know, I'm not scared to die…I think it taught me how to live in the present.

The camera angles relay various points of view. At first, the film is shot as though we are seeing through Bauby's left eye, with Bauby's internal thoughts voiced in the film, which is why it lends very well to dubbing. Bauby's memories are filmed in normal view but often without color, so we can tell it's a memory or flashback; there are also fantasies that are in full color. For a more realistic way to capture Bauby's view from one eye, the cinematographer filmed images in partial focus using a swing-and-tilt lens (Hartman 2007; Body and Being 2012). As one reviewer noted, the camera is Bauby and "shooting from Bauby's point of view has another advantage: It enables the film to avoid showing us what the paralyzed man looks like full-on for almost an hour" (Turan 2007).

The screenplay also features Sylvie de la Rochefoucauld, who was not as much a part of Bauby's life as Florence Ben Sadoun; in his memoir, she is mentioned only once as his children's "mother". Rochefoucauld's prominence in the film was later criticized by his circle of friends; she also appears more often than the character of Sadoun, when in reality it was the other way around. Sadoun noted several inaccuracies in the film, which were not in the book, including the wrong number of children; in the film, three children are shown rather than two (Schnabel admits this); she says she was also holding his hand when he died, which is not shown in the film. To correct the record, Sadoun wrote her own book entitled *La Fausse Veuve* ("The False Widow"—2008). The editor of *Elle*, who took over after Bauby, Valerie Toranian, was apparently so "offended by how Sadoun was portrayed that she did not allow filming at their offices and the magazine did not officially review the film." (Di Giovanni 2008). Sadoun and Bauby had met at *Elle* magazine in the early 1990s when Bauby was editor-in-chief and Ben Sadoun was a critic (Di Giovanni 2008).

Synopsis

This film captures the suddenness and inconvenience of a massive stroke on a very active man—Jean-Dominique Bauby—the editor of *Elle* magazine, based at the center of Paris fashion. It also captures the importance of communication as the only way to have any quality of life; the communication takes the form of blinking out letters that form words and an entire rich palette of a patient's experience. The film's portrayal of locked-in syndrome opens the film:

> The first 20 min are straight out of a nightmare. He emerges from a coma, facing a phalanx of nurses and doctors; he speaks, or thinks he does, only to realize no one can hear him; and he can't move a muscle. In one horrifying scene, he watches as his damaged but still-functioning right eye is sewn shut (Hartman 2008).

Another description of the opening:

> The movie begins claustrophobically, as we see the blurry bustle of the hospital room from Jean-Dominique's hazy, panicked perspective. Faces loom suddenly and awkwardly into view, while his captive consciousness writhes in its cage, trying to make contact with the world outside (Scott 2007).

The viewer has immediate access to Bauby's thoughts, narrated by the actor who plays him, Mathieu Amalric (unless watching a dubbed version in another language). Through this unique vantage point, we are also experiencing the neurological diagnosis of LIS and can see its limitations. The film takes us into his memories and fantasies, as well as through the process of his dictating his memoir. The activity of his mind drives the activity in the film. The film demonstrates the work of his speech therapist (played by Marie-Josee Croze), who helps him chiefly with communication. The therapist develops a communication code for Bauby using the alphabet reorganized according to the most commonly used letters in the French language. The therapist recites the alphabet and stops when he blinks; in a long process, this method provides the means for communicating words and sentences. The film is about "limitation, constraint, incarceration" (Scott 2007) but also about liberty through communication and an active mind.

The Social "Location, Location, Location"

The most striking aspect of the social location of this film is, in fact, the *location* of the hospital. This beautiful French Naval hospital, Berck-sur-Mer, is located in a coastal town of the same name on the French Channel coast, near Normandy where the film was shot on location. Location makes a difference when you only have one eye from which to see, and there is very little quality of life. This is where Bauby spent his last two years, post-stroke, and where he eventually died. Bauby writes of the hospital's location in his memoir as well, and although is clearly suffering the universal indignities of being a patient, we have to ask how much more he would suffer were he in a different type of hospital and location, where there may be more limited resources:

> The Naval Hospital must a be striking sight to the noisy light aircraft that buzz across the Berck shoreline at an altitude of three hundred feet.... The lighthouse and I remain in constant touch and I often call on it by having myself wheeled to ...the perpetually deserted terrace of Sorrel ward. Facing south, its vast balconies open onto a landscape heavy with the poetic and slightly offbeat charm of a movie set. (Bauby 1997: 27, 29.)

The location of the hospital speaks volumes about the French healthcare system overall, and the approach to patient rehabilitation in France, as Bauby clearly receives excellent rehabilitation care, and refers to his speech therapist as his "Guardian Angel": "She is the one who set up the communication code without which I would be cut off from the world... Speech therapy is an art that deserves to be more widely known." (Bauby 1997: 39–40.)

Bauby has the healthcare he does not because he is rich or famous, but because he is a French citizen enjoying the fruits of the nationalized French healthcare system. When the film was made, the French healthcare system was being ranked and studied; it was listed No. 1 by the World Health Organization in 2000. American viewers may actually marvel at the hospital and its location, and not believe it is realistic, and so it's important to note that *this is what stroke care looks like in France*. And it looks much different than it would for the average stroke patient in the United States.

Berck sur Mer began as a pediatric hospital for disabled children, and was expanded as a hospital for a variety of rehabilitation patients, including what Bauby, in his book, described as the "tourists" who are "survivors of sport, of the highway, and of every possible and imaginable kind of domestic accident"; essentially, trauma accident patients (Bauby 1997: 32). In fact, the hospital was so novel in 1913, it was the subject of a *New England Journal of Medicine* article (known then as *Boston Medical and Surgical Journal*) entitled: "The Marine Hospital at Berck-Srmer [sic], Principally for Crippled Children; The Largest Institution of Its Type in the World" (McMurtrie 1913). The article notes that the hospital's novel approach for "out of door treatment" for the crippled children of Paris is in a beautiful location with 14 miles of protected beach property. The article describes the marvelous facilities for rehabilitation patients, including warm fresh and saltwater baths, swimming pools, and many other features. Bauby, in his memoir, also notes some of the hospital's history.

The French Lifestyle

When teaching this film, it's also important to discuss Bauby's social location as a Parisian, and the French lifestyle, in which European social attitudes are far more liberal than Americans with respect to work-life balance, personal life, healthcare, the pursuit of leisure and pleasure, as well as governance, where there are many more political parties in France than the U.S. Paris in the mid-1990s was much different than today's Paris, which has now become Europe's "ground zero" for

Muslim extremism problems. In the 1990s, none of these issues were in the news, and Paris was at the center of fashion, food, and culture, and so was Bauby, in his role at *Elle*.

Paris fashion—Bauby's professional world—had been made famous by its "haute couture" (made to measure) approach throughout the early to mid 20th century—made famous in the U.S., in fact, through former first lady, Jacqueline Kennedy Onassis, who was noted for her "French clothes". By the 1970s, "pret a porter" (ready-to-wear) emerged more dominant, as French fashion began to be mass marketed, and there was a greater emphasis on marketing and manufacturing. By the 1990s, when Bauby was at the center of this world, there was a corporatization of many of the French couture houses, which became part of large multinational conglomerates (Dauncey 2003).

In the early 1990s, a fascination with the French diet and lifestyle began to dominate the news, coinciding with a now-famous *60 Minutes* piece entitled "The French Paradox" (airing November 17, 1991), in which veteran *60 Minutes* reporter, the late Morely Safer, explores why the rates of heart disease are so low in France given their high fat diets. In his report, he looked at why American males had a three times greater chance of dying of a heart attack than French males, even though they smoked, exercised less, and ate about 40 lbs. of cheese each year. (There are roughly 450 distinct types of French cheese grouped into eight categories.) Some mused in the report that cheese had beneficial health qualities and was therefore a "good dairy", whereas milk—consumed more by Americans—was the "bad dairy". A cardiologist noted in the report (dining with Safer at a French bistro in Lyon while looking at all the high fat foods on the menu):

> The farmers have been eating this for years. They've been eating a very high-fat diet, it seems, and yet they don't get heart disease; if we took the same diet and put it into an American, we would all be suffering from coronaries at an early age. There's something about the French that seems to be protecting them, and we're not sure what it is. We're looking for it (CBS 1991).

The report noted that the protection stemmed from the wine. Safer states: "The answer to the riddle, the explanation of the paradox, may lie in this inviting glass" (CBS 1991). The benefits of wine were related to drinking red wine in moderation. Meanwhile, in part of the United States known as "stroke alley" in the heart of the Bible Belt, the absence of wine was considered to be a risk factor for cardiovascular disease and stroke. After this report, wine sales in the United States dramatically increased.

More importantly, though, the report essentially covered the French attitude towards food as well as diet, and the difference in eating habits—the French took more care and pleasure with their food, eating more fresh foods, and tended to sit down at a table to eat and enjoy it, ate only three meals without snacks, while Americans were rushing, eating on the run, eating all the time, and eating more convenience or fast foods. This became a medical phenomenon at this time known as the "French Paradox," which I discuss further on (See History of Medicine).

The French have a saying, which is derived from the lyric of an old French torch song: *Regret nothing—in matters of love and food.* In short: The French are passionate about their food and really enjoy it. They never think of food as "sinful"; instead, they simply think of it as "tasty". Americans tend to think about food as either fuel or poison; they fear the effect the food will have on their bodies.

In 1995, France was also notable for being the home of the oldest living person in human history: On October 17, Jeanne Calment, a French woman who had outlived all of her grandchildren, turned 120; she died at 122 in 1997, the same year as Bauby. She was being studied at the time to see if her diet or lifestyle was remarkable in some way (Whitney 1997).

Thus, Bauby's stroke at 43—in 1995—was less typical in France than the U.S., given the statistics of cardiovascular events for French males at the time. Bauby details in his memoir that his stroke occurs on December 8, 1995, and discusses his last normal day as fairly stressed; this includes reference to a general strike in Paris that has paralyzed the city's transit system, making it almost impossible to travel from point A to point B; Bauby also makes a slight reference to the health of the recently retired French President, Francoise Mitterrand.

Bauby's last normal day was an historic time capsule in 1995 France. He is referring to the Paris Strikes that occurred in November through the end of December in 1995, which was a general strike in the public sector mounted in protest to proposed government cutbacks to social benefits. It was considered the largest social resistance in Paris since 1968 (Trat 1996). In May 1995, after Francois Mitterrand left office—he died in 1996 after it was discovered that he had hidden his advanced prostate cancer from the country through falsified medical records—(Whitney 1996)—a right-wing, conservative President had been elected (Jacques Chirac); his prime minister, Alain Juppe, proposed substantial cuts to social programs, affecting a range of workers and retirement benefits. Protests began in October, major strikes and demonstrations carried on until December 15, 1995—about a week after Bauby's stroke.

But when the film premiered in 2007, it coincided with the release of Michael Moore's *Sicko* (see Healthcare Ethics on Film), which spent considerable time explaining the superiority of the French healthcare system, but also the French political infrastructure in which "protests" were common, and which helped to create, and keep, a very progressive government where citizens enjoyed numerous social benefits as part of the French lifestyle. In fact, Charles De Galle famously stated: "How can you govern a country which has 500 varieties of cheese?" Thus, when considering the equivalent of a "wine pairing" of this film, showing it with *Sicko* may work well.

History of Medicine

There are three history of medicine contexts relevant to this film. Primarily, a discussion of ischemic stroke and locked-in syndrome (LIS) is key, and it may be prudent to time this film for medical students with a curriculum that focuses on neurological disorders. Second, the history of the Speech Language Pathology

(SLP) field is prudent. If planning an ethics curriculum for SLP trainees, this would be the film to show, over any others in this book's collection. Third, this film demands a discussion of the French healthcare system in general, which lies in contrast to the American healthcare system—years away from providing universal health coverage to its citizens. However, for that topic, please refer to Healthcare Ethics on Film. Finally, this film also prompts an epidemiological review of the literature on the "French Paradox" which proposes that cardiovascular disease is lower in France than the U.S. because of key factors in diet and lifestyle.

Locked in Syndrome (LIS)

LIS is a more recent medical phenomenon that was first described in 1966 by Plum and Posner as: "quadriplegia, lower cranial nerve paralysis, and mutism with preservation of consciousness, vertical gaze, and upper eyelid movement." This was defined as a false coma coined 'locked-in syndrome' Twenty years later (1986) LIS was redefined as "quadriplegia and anarthria (inability to articulate words) with preservation of consciousness." (Smith and Delargy 2005).

In 1995, the American Congress of Rehabilitation Medicine defined LIS as "a syndrome characterized by preserved awareness, relatively intact cognitive functions, and by the ability to communicate while being paralyzed and voiceless." (NARIC 2013.)

LIS results from a brainstem lesion following damage to a part of the brainstem called the pons. This can occur during an ischemic stroke where there is a lack of blood flow due to a blood clot, or when there is a hemorrhagic stroke (bleeding in or around the brain). Other causes of LIS include trauma. In Bauby's case, he most likely suffered an ischemic stroke, accounting for 90% of all strokes; in his memoir he relays that his physicians call it a "cerebrovascular accident that took my brain stem out of action" (Bauby 1997). LIS is typically caused by an "ischemic pontine lesion" (Smith 2005). People with LIS have the most severe form of quadriplegia as they typically have no movement other than their eyes; they can't control their breathing; their mouths or face, which means they can't swallow or speak, but their consciousness and all cortical functions remain. Thus, quality of life for LIS patients is completely dependent upon facilitating communication, in addition to round-the-clock nursing care. Unfortunately, many LIS patients in the past were not typically recognized as having LIS unless they showed obvious signs such as blinking at critical moments. A key French paper in the literature on LIS entitled "Blink and You Live" (Bruno et al. 2008) points out that "more than half of the time physicians fail to recognize early signs of awareness in LIS." The paper makes clear that LIS patients should be given every opportunity to live, but not denied the opportunity to die with withdrawn life support. With good medical care, LIS patients can live several decades in their states, but some will succumb to complications, as in Bauby's case, who died from pneumonia (likely aspiration pneumonia) two days after the publication of his memoir. In the 1990s, when Bauby suffered from LIS, blinking to an alphabet code was the most sophisticated tool available for communication. By the time the film was released in 2007, eye-controlled computer technology was available for LIS patients, which has improved communication and facilitated more control of their environments. Thus, for LIS, facilitating communication is the only real goal of care, as there is really no "treatment" other than daily nursing care. Most LIS patients will need support for breathing and feeding (Bruno et al. 2008).

No one actually knows the real prevalence of LIS, since many in this state will die shortly after the stroke, but the literature currently states that less than 1% of all stroke patients will have LIS; but since half are never recognized, it remains an ethically problematic diagnostic dilemma. Most patients with LIS are younger (17–52); Bauby was thus representative of the patient population. *Eye movement is the basis for communication,* but often sight is impaired as well. LIS is considered to be the most severe neurological condition in a hospital setting because of the need for specialized and dedicated nursing care, respiratory support through tracheostomy and gastronomy tube feeding. Patients with LIS can only be considered "capacitated" to make decisions if they are "recognized" and communication is facilitated. For this reason, the critical healthcare provider in LIS is a skilled speech language pathologist (SLP), and not all facilities will have this resource available to LIS patients. In fact, I have had clinical ethics cases in which I have found insufficient tools made available to LIS patients, and thus, include SLP consultations as part of my recommendations to restore decision-making capacity. In cases of LIS, since all language skills are preserved, communication is based on facilitating patients to spell out words using alphabet codes. As in Bauby's case, patients would indicate the desired letter by moving their eyes or blinking to the right letter, and the letters do not have to correspond to the alphabet in order—they may be grouped according to the most used letters (as Bauby's SLP arranged), or short messages and phrases can be added so that blinking can be to a particular short phrase (Laureys et al. 2005; Rousseau et al. 2015). More recently, computer technology has been used to facilitate communication, but it depends upon resource allocation. In a 2003 review of a large cohort of LIS patients, the majority had some quality of life and did not consider withdrawal of life support; out of those who considered withdrawal of life support, they all decided to be full code once enhanced communication tools were made fully available (Laureys et al. 2005). In these patients, being able to communicate seems to renew a strong desire to live even within their limited states, which may have to do with a completely preserved personhood. Of note, LIS patients seem to have a great desire for social participation; many use their communication tools to reach out to the public to voice their experiences. Bauby's memoir is thus considered now to be a more typical project than atypical in LIS patients.

The History of Speech Language Pathology and Communication Disorders

When teaching this film, the "star" of the healthcare team is the speech language pathologist (SLP) who is responsible for unlocking the patient's ability to communicate. The role of the neurologist is diminished to diagnosis, but there is no treatment available. It's important then, to cover (at least briefly) the history of the SLP field (Duchan 2011), which is traced to eighteenth century England with the field of "elocution", meaning speech perfection. The elocutionists would work with their clients to help them with oration, pronunciation, and diction—dramatized in the relationship of Henry Higgins to Eliza Doolittle in the play, "Pygmalion" (later turned into the Broadway play and film, *My Fair Lady*). A key figure in the history of SLP in the nineteenth century was Alexander Graham Bell, in his work with the deaf, inspired by his own deaf mother. Bell studied the physiology of speech, and began to teach speech to deaf students using a universal alphabet invented by his father called "Visible Speech"—distinct from American Sign Language. In 1872 Bell opened a school in Boston to train teachers of deaf children. Other famous pioneers in the field of SLP included Annie Sullivan, who was taught by Laura Bridgeman (the first deaf-blind individual to be educated); Bridgeman learned the manual alphabet invented in France by Charles-Michel de l'Eppe (1712–1789) at the Perkins School for the Blind, and taught it to Sullivan. Sullivan then later worked with Helen Keller, and established herself as a pioneering speech language pathologist. For space considerations, I do not go into the body of work of l'Eppe, Bell, Bridgeman or Sullivan, but depending on how you're using this film, it may make for interesting supplementary material to expand on the spectrum of the SLP field, and the globally famous figures in the field. The National Association of Elocutionists was founded in 1892, and began to become a recognized profession that has now been absorbed into the broader field of SLP.

A major turning point in the field of SLP was a book by physician, Samuel Potter who suffered from severe stuttering, and wrote a book entitled Speech and its defects (Potter 1882; UNC 2016). It focuses on stuttering (dyslalia), but also provided a comprehensive review of a variety of communication disorders, and suggested remedies. In 1926, the American Academy of Speech Correction was established, and post-WWII, there was a particular focus on treating aphasia associated with brain injuries. The field expanded throughout the 1950s on how the brain processed language and speech. When testing and other technologies began to evolve, there was more understanding about the development of language and speech, and delays or impediments.

The alphabet code approach used in the film is also known as "partner assisted scanning", which is essentially a form of communication that requires a partner to assist the person with a communication disorder, such as patients with LIS. Depending on the curriculum or audience for this film, pairing more information about this specific SLP technique with scenes in the film that demonstrate it, may be useful.

Cardiovascular Disease and the French Paradox

As discussed earlier, American fascination with the French lifestyle was based on dramatically different statistics in cardiovascular disease among the two countries. In the United States, as heart disease rates began to increase dramatically in the

1970s and 1980s (see discussion in Chap. 4), French researchers wondered why, with the French diet being so high in saturated fat, were the rates of heart disease among the French so much lower than in the British and North Americans—whose diets were high in saturated fats? Dubbed "The French Paradox," investigators began to look more closely at the French diet, a diet that seemed to be filled with "bad fats"—creams, butter, frois gras—all saturated fats, saturated in even *more* saturated fats. But then there was a key difference: the French drank red wine with their meals. Studies began to look at whether moderate drinking—such as red wine sipped with meals—contributed to lower rates of heart disease, and found that it did, but only in moderation. The Mediterranean diets are also based on regular yet moderate consumption of red wine. Ultimately, we later learned that the French Paradox involved other factors, such as differences in lifestyle (Rosenthal 2004). When news of the French Paradox and red-wine drinking was broadcast on *60 Minutes* in November, 1991 (CBS 1991), sales of red wine skyrocketed. But studies were not able to show, conclusively, that moderate wine drinking was any better than moderate intake of any alcohol. The wine industry began funding more research into studies that looked at specific properties of red wine, and how these might affect blood cholesterol and overall rates of heart disease. Much of this research supported that moderate drinking—defined as two drinks per day for men, and one drink per day for women—was not harmful, and potentially healthful in that it was associated with lowering cholesterol levels. The American 1990 guidelines regarding alcohol consumption stated: "Drinking has no benefits" and did not recommend any alcohol. By 1995, U.S. alcoholic guidelines read: "Alcoholic beverages have been used to enhance the enjoyment of meals by many societies throughout human history." No health claim was made, but nonetheless, it was inferred. The wine industry was now anxious to begin promoting wine as a health food for its own gain. But because moderate drinking can so easily cross into alcohol abuse, health claims about red wine are difficult to make. By 2000, the U.S. guidelines made it clear that excess alcohol consumption was harmful; this made it impossible for the wine industry to quote guidelines on their labels. Few people truly understand what the studies really show about red wine and cholesterol (Rosenthal 2004).

As early as 1979, French researchers demonstrated a relationship between low rates of heart disease and wine, in spite of diets rich in saturated fats. The most comprehensive study that appears to have solved "The French Paradox" was not a wine study per se, but an international study looking at heart disease rates around the world, including rates of heart disease in France and Mediterranean countries. Known as the WHO MONICA Project (The World Health Organization Monitoring of Trends and Determinants in Cardiovascular Disease), this was a 10-year study that monitored deaths from heart disease and risk factors in men and women 35–64 in various communities between 1985 and 1987; it was inspired by Keys Seven Countries Study (see Chap. 4). This study found major differences in incidence of heart disease and death rates, a surprising finding, given the fact that there were even differences in regions with common diets, such as those of France, Britain and the United States. MONICA was the study that began The French Paradox; it also confirmed data from the Seven Countries Studies regarding low rates of heart disease in Mediterranean countries. MONICA led some investigators to look at the properties in olive oil and garlic while others looked at properties in wine (Rosenthal 2004).

Throughout the 1990s, more convincing evidence regarding red wine began to appear. The Copenhagen City Heart Study was another good study looking at wine and heart disease. Published in 1998, this study followed wine drinkers for 10–12 years and found that moderate wine drinking, rather than beer or spirits drinking, was associated with a lower risk of stroke. As a result of red wine studies, we now know that red wine is believed to have a heart healthy effect in three ways: it is antioxidant, vasodilating, and antithrombotic (Rosenthal 2004).

Statistical errors in analysis have also been posited as accounting for a seeming "French Paradox". Britain and the U.S., which were compared with France, had been consuming red meat in large quantities for longer periods than in France. Saturated fats from meats are different than saturated fats from cheeses or other dairy products. It was not until the 1970s, that France's saturated fats consumption started to truly compare to Britain and the U.S. Prior to 1970, the French were consuming a diet of roughly 21% of calories from fat—which is a relatively low fat diet. Some researchers proposed that in order to do a truly comparative study, we would have to wait to see what the effects of the 1980s diet had on France's 50 + generation. The heart disease rates in France found by the MONICA study were likely based on the pre-1970s diet, in which 21% of calories were from fat. Other explanations for the French paradox were that not all deaths from heart disease were accurately recorded as being caused by "heart disease". Instead, complications of heart disease were often cited as the cause of death, which affects the data. It's also noted that in countries where wine drinking is prevalent, other drinks are not. Wine is cheaper in many European countries than water or soft drinks or juice; in addition, France, Italy, and Spain all shared low consumption of saturated fat until recently (Rosenthal 2004).

But if we were to repeat these studies today, heart disease deaths have decreased due to better medications for hypertension and high cholesterol. By 1993, for example, 34% of survivors of heart attacks in France took cholesterol lowering drugs compared with 4% of survivors in Britain; 63% of French took aspirin compared with 38% in Britain; another 20% took anticoagulants compared with 5% in Britain. Finally, another factor in the French Paradox is that obesity rates are lower because fast food has not infiltrated France as much as it has in the U.S (Rosenthal 2004).

Clinical Ethics Issues

While one can discuss a variety of clinical ethics issues in this film, ranging from autonomy and personhood to beneficent care plans, or even the concept of withdrawal of life support (see Chap. 2), I would hone and focus the clinical ethics discussion here to decision-making capacity, its criteria and parameters. When I teach this film, I provide a review of decision-making capacity as a concept housed under the Principle of Autonomy or Respect for Persons (in the latter, there is an obligation to protect non-autonomous patients if they don't have capacity; in the former, to respect autonomous patients wishes regarding self-determination). Thus, students need to understand the distinctions between competency and capacity; and the "U-ARE" criteria, an acronym for Understanding, Appreciation, Rationality and Expression (of a choice, preference or decision). Patients need to satisfy all these criteria in order to be considered decisional. Thus, all three of the films discussed in this section emphasize different aspects of the U-ARE criteria as barriers to decision-making capacity. This film serves as a laboratory in which to examine *Expression* as the chief barrier to decision-making capacity.

Competency, Capacity and Communication

For a much more detailed exploration of competence, see Chap. 6, which discusses the psychiatry ethics film, *One Flew Over the Cuckoo's Nest.* Briefly, it's important to note that competency is a *legal standard* used to determine legal guardianship; while decision-making capacity is a medical standard for decision-making based on the medical judgment of the Attending physician in any field, which determines whether a surrogate decision-maker is needed for a particular medical decision, which is not the same thing as a *court-appointed guardian.* Typically, neurologists and psychiatrists are frequently asked to assess underlying functional or mental health barriers to decision-making capacity, while clinical ethicists can frequently assess whether patients have enough information to meet Understanding and Appreciation with respect to informed consent. But at times, clinical ethicists will also need to point out barriers to expression and communication interfering with capacity. For example, common clinical ethics consults may involve intubated, ventilator-dependent patients who appear unwilling or unable to make decisions; in many such cases, SLP consultations can facilitate better tools to communicate, or different types of ventilators in which speech is easier. Sometimes clinical ethicists will point out missed problems with hearing impaired patients who may "mask" their impairment—particularly if they are suffering from drug-related tinnitus, or are in an ICU for a completely different reason and no one thinks to check their hearing when they make irrational statements (because they misheard). Patients who don't speak the same language as their practitioners may not readily volunteer they don't understand, while many patients are not provided with interpreters. Clinical ethicists have also seen properly diagnosed LIS patients who still have inadequate communication tools at the bedside.

   In this film, the patient is already under the care of a neurologist for his brain stem injury, and he is properly diagnosed with LIS. But based on the LIS literature, many patients are often misdiagnosed. So it would be important to explore the existential suffering of patients who are neurologically misdiagnosed, and how simply facilitating communication completely enables autonomy.

The U-ARE Criteria

The U-ARE criteria was presented by Roth et al. (1977), and established the necessary components for decision-making capacity. (See Table 5.1.) This is the same criteria used in any formal decision-making capacity assessment used in any mental health consultation; the criteria should be used by clinical ethicists as well

Table 5.1 Criteria for decision-making capacity

Distinct from competency (which is a legal determination, made by a judge for decision-making authority), decision-making capacity is a medical determination that can be made by the Attending or responsible physician; all competent persons have presumption of capacity. Decision making capacity can change over short periods of time, and be affected by physiological changes

The criteria to assess capacity is U-ARE:

- Understanding: does the patient understand his/her medical status, treatment options, and risks associated with the various options? Is there a basic understanding of the facts involved in that decision? Are there language or literacy barriers?
- Appreciation: does the patient have some appreciation of the nature and significance of the decision?
- Rationality (or ability to Reason): is the patient able to reason with the information in order to make a decision? Can the patient engage in reasoning and manipulate information rationally? Does the patient make rational decisions, or offer a rationale for a seemingly irrational decision, based on preferences, beliefs, culture, and so forth? In patients with dementia, or mental health status changes, look for decisional stability over time, in contrast vacillation, which indicates an absence of capacity
- Expression of a Choice: Can the patient communicate a decision, or express him/herself in some way (orally, in written form, in gestures, or blinking)? This is important for patients with aphasia or who cannot speak

Some patients may require a more formal capacity assessment, which may need to involve a psychiatric consultation or an ethics consultation

Standards for Capacity

Decision making capacity relates to a particular decision at a particular time; may exist for simple but not complex decisions; and may exist for medical but not other decisions

Decision-making capacity operates on a sliding scale that permits lesser standards of capacity for less consequential medical decisions (such as getting a flu shot), and requires higher standards of capacity for greater consequential decisions such as consenting to radioactive iodine or total thyroidectomy. The more serious the expected harm to the patient from acting on a choice, the higher should be the standard of decision making capacity. However, no single standard for capacity is adequate for all decisions. The standard of capacity necessary depends on the risk involved, and varies from low to high

Sources

1. Roth, Loren H., Alan Meisel, and Charles W. Lidz. 1977. Tests of competency to consent to treatment. American Journal of Psychiatry 134:279–284
2. Etchells Edward, Gilbert Sharpe, Carl Elliott, and Peter Singer. 1996. Bioethics for clinicians:
3. Capacity. Canadian Medical Association Journal 155:657–661
4. Buchanan, Alan E., and Dan W. Brock, 1989. *Deciding for Others: The Ethics of Surrogate Decision Making.* Cambridge: Cambridge University Press
5. Grisso, Tom and Paul A. Appelbaum. 1998. *The Assessment of Decision-Making Capacity: A Guide for Physicians and Other Health Professionals.* Oxford: Oxford University Press
6. Culver, Charles, M., and Bernard Gert. 1990. The Inadequacy of Competence. Milbank Quarterly 68: 619–643

who are asked to weigh in on decision-making capacity. Clinical ethicists will typically recommend a formal assessment by a psychiatrist or other mental health assessor first to rule out mental health barriers to capacity. In cases where neurological injury is suspected, a neurology consult will determine functional problems with capacity. Otherwise, many barriers to capacity are based on communication problems having to do with medical literacy or language barriers. In these cases, Understanding and/or Expression can be impaired anytime a patient is functionally unable to hear or speak. Restoring decision-making capacity in these cases are easily solved by enabling the right tools of communication. For LIS patients, alphabet codes for blinking can restore decision-making capacity, but there will still be a need for a surrogate decision-maker to act as the patient's legal voice: signing documents, for example, or voicing the words the patient has indicated through blinking. In cases of hearing impaired patients, enabling writing tools is important, or deaf interpreters for sign language. In cases of language barriers, patients require interpreters as well; the common faux pas of relying on patients' family members to interpret may compromise voluntariness (or even privacy rules in some cases), and thus, professional interpreters are required to meet ethical and legal standards. Ultimately, we can clearly see in this film that although Bauby completely understands and appreciates his condition, and is able to rationally manipulate the information to make medical decisions consistent with his values and preferences (the standard for Rationality), if he can't express himself, he lacks capacity. In the first scene of the film, for example, Bauby's thoughts are voiced to the viewer, in which he is refusing his eye surgery, but yet he isn't yet enabled to express his refusal. It is important to convey we typically demand all the U-ARE criteria to be met for high stakes decisions. I expand on this criteria more in Chap. 7 on *Still Alice*, which deals with the nuances of loss of capacity in various stages of dementia that affect every aspect of the U-ARE standard.

Conclusions

Diving Bell and the Butterfly is the perfect film to use when examining decision-making capacity because it demonstrates one of the most overlooked barriers: expression and communication when there are high levels of understanding and appreciation. By extension, when communication is enabled or facilitated, autonomy is unmasked or restored. It is important to ensure that there is a clear understanding of the distinctions between LIS and persistent vegetative state, as in the latter autonomy is disabled, and thus personhood is not a factor. Finally, discussions about the extent of existential suffering in LIS or worse, unrecognized LIS, can be important, but one can become mired if there is more focus on the suffering and less focus on how enabling communication in a practical way can create quality of life. As noted, there may be different audiences for this film depending on whether it is used for medical school trainees, rehabilitation medicine trainees; specifically SLP trainees, or even as a narrative medicine film focusing more on literature. At its heart, communication as an important feature of what it means to be fully alive and a person is the focus for any audience.

Film Stats and Trivia

- Ranks 77 in the Top 100 films of the 21st century by the BBC.
- Schnabel won best director at the 2007 Cannes Film Festival and Best Director at the 65th Golden Globe Awards, where the film also won Best Foreign Language Film.
- Since it was produced by an American company, it was ineligible for the Academy Award for Best Foreign Language Film.
- Actor Max von Sydow plays the father in a flashback scene.

Theatrical Poster

Director: Julian Schnabel
Adapted Screenplay: Ronald Harwood
Editing: Juliette Welfling
Cinematography: Janusz Kaminski
Based on the book "Le Scaphandre et le Papillon" by Jean-Dominique Bauby;
Editor: Juliette Welfling
Music by Paul Cantelon
Producers by Kathleen Kennedy and Jon Kilik
Released by Miramax Films
Running time: 1 h 52 min
Starring: Mathieu Amalric (Jean-Dominique Bauby), Emmanuelle Seigner (Céline), Marie-Josée Croze (Henriette), Anne Consigny (Claude), Patrick Chesnais (Dr. Lepage).

References

Bauby, Jean Dominique. 1997. *The Diving Bell and the Butterfly*, trans. J. Leggatt. New York: Vintage Books.
Beineix, Jean-Jacques. 1997. *Assigned Residence*. Cargo Films, France, released March 14.
Body and Being. 2012. Film techniques in the Diving Bell and the Butterfly. Body and Being. http://bodyandbeing.lmc.gatech.edu/bab_wiki/index.php/Film_Techniques_in_The_Diving_Bell_and_the_Butterfly. Accessed 23 Feb 2018.

Bruno, M.A., F. Pellas, C. Schnakers, P. Van Eeckhout, J. Bernheim, K.H. Pantke, F. Damas, M. E. Fayomonville, G. Moonen, S. Goldman, and S. Laureys. 2008. Blink and you live: The locked-in syndrome. *Rev Neurol (Paris)* 164:322–335. https://www.ncbi.nlm.nih.gov/pubmed/18439924.

CBS. 1991. The French Paradox. *60 Minutes*, November 7. http://www.cbsnews.com/news/how-morley-safer-convinced-americans-to-drink-more-wine/.

Dauncey, Hugh (ed.). 2003. *French Popular Culture: An Introduction*. London: Arnold Publishers.

Di Giovanni, Janine. 2008. The real love story behind The Diving Bell and the Butterly. *The Guardian*, November 29. https://www.theguardian.com/lifeandstyle/2008/nov/30/diving-bell-butterfly-florence-bensadoun.

Duchan, J.F. 2011. A History of Speech-Language Pathology. University of Buffalo Judy Duchan Webpage. http://www.acsu.buffalo.edu/~duchan/new_history/overview.html. Accessed 23 Feb 2018.

Dumas, Alexander. 1996. *The Count of Monte Cristo*, trans. R. Buss. London: Penguin Books.

Hartman, Darrell. 2007. Schnabel's Portrait of an Artist in Still Life. *The Sun*, September 28. http://www.nysun.com/arts/schnabels-portrait-of-an-artist-in-still-life/?page_no=2.

Laureys, S., F. Pellas, P. Van Eeckhout, S. Ghorbel, C. Schnakers, F. Perrin, J. Berre, M.D. Faymonville, K.H. Pantke, F. Dams, M. Lamy, G. Moonen, and S. Goldman. 2005. The locked-in syndrome: What is it like to be conscious but paralyzed and voiceless? *Progress in Brain Research* 150:495–511. https://www.ncbi.nlm.nih.gov/pubmed/16186044.

McMurtrie, D.C. 1913. The Marine Hospital at Berck-Surmer, principally for crippled children; the largest institution of its type in the world. *The Boston Medical Surgical Journal* 168:14–16. http://www.nejm.org/doi/full/10.1056/NEJM191301021680104.

National Rehabilitation Information Center (NARIC). 2013. What is locked-in syndrome? http://www.naric.com/?q=en/FAQ/what-locked-syndrome. Accessed 23 Feb 2018.

Potter, Samuel. 1882. *Speech and Its Defects: Considered Physiologically, Pathologically, Historically and Remedially*. Philadelphia: P. Blakiston, Son & Co.

Rosenthal, M. Sara. 2004. The Olive and the Vine: The Mediterranean Diet and The French Paradox. In: *The Skinny on Fat: A Look at Low Fat Culture*, Rosenthal, M. Sara, 112–126. Toronto: Macmillan Canada.

Roth, Loren H., Alan Meisel, and Charles W. Lidz. 1977. Tests of competency to consent to treatment. *American Journal of Psychiatry* 134: 279–284.

Rousseau, M.C., K. Baumstarck, M. Alessandrini, V. Blandin, T.B. de Villemeur, and P. Auquier. 2015. Quality of life in patients with locked-in syndrome: Evolution over a 6-year period. *Orphanet Journal of Rare Diseases* 10: 88. https://www.ncbi.nlm.nih.gov/pmc/articles/PMC4506615/.

Scott, A.O. 2007. Body unwilling, a mind takes flight. *New York Times*, November 30. http://www.nytimes.com/2007/11/30/movies/30divi.html.

Smith, E., and M. Delargy. 2005. Locked-in syndrome. *British Medical Journal* 330: 406–409. https://www.ncbi.nlm.nih.gov/pmc/articles/PMC549115/.

Stock, Francine. 2008. Interview with Julian Schnabel. *The Guardian*, February 8. https://www.theguardian.com/film/2008/feb/08/guardianinterviewsatbfisouthbank.

Trat, J. 1996. Autumn 1995: A social storm blows over France. *Social Politics* 3: 223–236. https://doi.org/10.1093/sp/3.2-3.223.

Turan, Kenneth. 2007. Through his eyes: "The Diving Bell" imbues a unique plight with an uplifting, even funny perspective. *Los Angeles Times*, November 30. http://articles.latimes.com/2007/nov/30/entertainment/et-diving30.

Uhlich, Keith. 2007. On the Circuit: The Diving Bell and the Butterfly. *Slant Magazine*, August 21. http://www.slantmagazine.com/house/article/on-the-circuit-the-diving-bell-and-the-butterfly-le-scaphandre-et-le-papillon.

University of North Carolina. 2016. History of the professions: A brief history of speech-language pathology. https://hsl.lib.unc.edu/speechandhearing/professionshistory. Accessed 23 Feb 2018.

Whitney, Craig R. 1996. Francois Mitterrand dies at 79; champion of Europe. *New York Times*, January 9. http://www.nytimes.com/1996/01/09/world/francois-mitterrand-dies-at-79-champion-of-a-unified-europe.html.

Whitney, Craig R. 1997. Jeanne Calment, world's elder, dies at 122. *New York Times*, August 5. http://www.nytimes.com/1997/08/05/world/jeanne-calment-world-s-elder-dies-at-122.html.

Williams, A.N. 2003. Cerebrovascular disease in Dumas' The Count of Monte Cristo. *Journal of the Royal Society of Medicine* 96: 412–414. https://www.ncbi.nlm.nih.gov/pmc/articles/PMC539579/.

Competency and Psychiatry Ethics: One Flew Over the Cuckoo's Nest (1975)

6

When you ask psychiatrists about films they find useful in teaching about their profession, *One Flew Over the Cuckoo's Nest*, which won several academy awards, including Best Picture, is still a classic that makes their list. Based on Kesey's (1962) novel of the same name, this film is about a prisoner malingering as "insane" who is sent to a psychiatric hospital in Oregon (where the film was shot) to be assessed for competency. Things don't go well, and the prisoner, Randle McMurphy, winds up having involuntary electroconvulsive therapy (ECT), and in the end, is also lobotomized, which was not an uncommon practice in 1962, when the novel was published. This film also helps to explain current psychiatric medico-legal issues by demonstrating past psychiatric abuses. From a clinical ethics standpoint, this film is a perfect vehicle for deeper discussions about competency as a legal standard, which is distinct from decision-making capacity, a medical determination limited to a particular decision. This film is also a vehicle to discuss clinical ethics issues that arise for prisoner patients, as well as the dilemma of voluntary vs. involuntary treatment, and ethical problems arising from untreated behavioral health problems in hospitalized patients. Finally, this film is useful as a History of Medicine film to discuss the history of psychiatry, as this film is considered a realistic picture of what institutionalization looked like in the 1960s, before the closing of institutions after the *Community Mental Health Act* was passed in 1963 and took effect. Of note, the medical director of the Oregon State Hospital, where *One Flew Over the Cuckoo's Nest* was filmed on location, plays himself in the film, which is a testament to its realism. Since this film was first released in 1975, it is still screened at current psychiatric conferences, and is recommended to psychiatric residents. This film is considered "culturally, historically, or aesthetically significant" by the United States Library of Congress, and was selected in 1993 for preservation in the National Film Registry (Library of Congress 1994).

© Springer International Publishing AG, part of Springer Nature 2018
M. S. Rosenthal, *Clinical Ethics on Film*,
https://doi.org/10.1007/978-3-319-90374-3_6

The Cuckoo Clock: Origins and Timelines of *One Flew Over the Cuckoo's Nest*

One Flew Over the Cuckoo's Nest originated from a 1962 fictional novel of the same name by Ken Kesey (1935–2001), who was part of the "Beat Generation" (see Chap. 4, on *All That Jazz.*) The book was informed by his personal experiences as a research subject in a Veteran's Administration (VA) hospital study on hallucinogenic drugs (LSD), as well as working as an attendant on a psychiatric ward of that same VA hospital (Biography.com 2017; Oregon PBS 2005; Rich 2012). Kesey became a strong advocate for widespread use of hallucinogenic drugs as a vehicle for advancing "spiritual awakening" and became better known as an LSD promoter during the height of the Hippie movement. Inspired by his own Beat generation's Jack Kerouac, author of On the Road (1958), he, too, went "on the road" with his friends (dubbed "The Merry Pranksters") in 1964 in a big school bus (called "Further") decorated with psychedelic graffiti. This psychedelic road trip coincided with Kesey's second book, A Great Notion (1964). While travelling, Kesey met up with both Kerouac, and another LSD proponent, Timothy Leary, whose famous adage was: "Tune in, turn on and drop out." Kesey's cross-country road trip involved stopping at various places and conducting "Acid tests"—which were essentially LSD parties (or a "communal trip") where, for one dollar, a partygoer would get LSD-laced Kool Aid, and listen to live music by Kesey's friends (often the band, The Grateful Dead) and have an LSD trip. Kesey essentially collected a following, and is noted more for being one of the early figures associated with the 1960s counterculture (a.k.a. "Psychedelic Sixties") than he is as the author of One Flew Over the Cuckoo's Nest (Kesey 1962). Kesey was arrested and sent to jail in 1967 on drug possession charges, and he and his Merry Prankster road trips is the subject of Tom Wolf's book, The Electric Kool-Aid Acid Test (Wolf 1968). The book is considered a sociological core text chronicling this period of the 1960s; it also details Kesey's exile to Mexico and his arrests. Eventually, Kesey dies from liver failure (presumably due to his overt and frequent drug use) in 2001. Kesey is responsible for the oft-used phrase "drinking the Kool Aid"; he used it first before the infamous Jonestown mass suicide imitated lacing chemicals in a drink (Jonestown members drank a different brand—Flavor Aid (Higgins 2012). Ultimately, the term "drinking the Kool Aid" became synonymous with following a cult of personality.

Ironically, Kesey, like his 1962 character, McMurphy, served a six-month sentence on a work farm. The documentary, *Magic Trip: Ken Kesey's Search for a Kool Place* (2011) was about this period in Kesey's life. LSD became an important part of Kesey's life after One Flew Over the Cuckoo's Nest was published; he often wrote under the influence of LSD, used it continuously as a recreational drug, which some could argue did not ultimately serve him well from an addiction health perspective. Kesey is an example of a writer who peaked early, but had difficulty living up to his earlier works, and it is unclear whether he would be framed today as someone who was living under the influence of drug addiction. Similar to his peer,

Bob Fosse (see Chap. 4), he was a product of a generation that saw drug use as a hedonistic, consciousness-expanding tool, rather than in the current prism we would construct as addiction.

1962: The Book

In 1962, 27 year-old Kesey published <u>One Flew Over the Cuckoo's Nest</u>, which he began writing in 1959. It has since been listed as one of the 100 best novels of the century (Grossman 2010). The book raises classic questions associated with the Beat Generation surrounding rebellion, conformity, role of authority on personal freedoms and autonomy, as well as unconscious social controls. Kesey's cast of characters, and the realistic structure described, are informed by his own interactions with patients while he was an aide/orderly on the psychiatric wards of the Palo Alto Veterans Administration Hospital, affiliated with Stanford University, where Kesey was a student (Oregon PBS 2005). Although not highlighted in the film, the novel continuously makes clear that the cast of characters on the ward are veterans. This work serves as an important time capsule in the history of psychiatry as it is just prior to de-institutionalization reforms in mental health (see further under History of Medicine). The book's protagonist is Randle Patrick McMurphy, a convict guilty of pettier crimes: battery and gambling, as well as a charge of statutory rape. The novel is situated in Kesey's home location of Oregon; in order to escape a prison work farm, McMurphy fakes insanity, thinking the psychiatric ward will be a lighter sentence; in fact it is a much worse environment that ends his life. The novel's narrator is partly Native American, Chief Bromden, a large, muscular man who feigns to be deaf-mute, which allows him to maintain greater autonomy and privacy, and also become an informant of his environment. Bromden, a WWII veteran, is also a victim of cultural abuses, and his moral distress leaves him traumatized, with a diagnosis of schizophrenia. Bromden refers to the "Combine" as an organized system of government control over society and individual autonomy. Kesey creates Bromden as an unsung hero of a growing Civil Rights movement, in which Native Americans, the first casualties of White Supremacy, are in fact "muted" from the larger discourse on human rights; it is not until later in the 1960s, when Native Americans begin to visibly organize around human rights abuses and their disenfranchisement (see under Social Location). Unlike the film, the book explains the title: Chief Bromden, after receiving ECT, recalls the following nursery rhyme attributed to Oliver Goldsmith of *Mother Goose* fame (Sutherland 1990).

Vintery, mintery, cutery, corn,

Apple seed and apple thorn;

Wire, briar, limber lock,

Three geese in a flock.

One flew east,

And one flew west,

And one flew over the cuckoo's nest.

An important feature of the book is the use of psychiatric medications as mind and behavior control (chemical restraint). While Kesey was a student at Stanford University, he volunteered as a human subject in a large government drug study known as Project MKUltra, (U.S. Senate 1977) in which subjects were given hallucinogenic substances to test limits and parameters of psychological control. (This was roughly a decade prior to the infamous Stanford Prison Study, discussed further on under History of Medicine). Social control of the inmates is also achieved by Nurse Ratched through a system of rewards, praise and public shaming.

1963: The Play

The actor, Kirk Douglas, read a pre-publication galleys copy of the novel in 1961 (Kyselyak 1997) and immediately sought to purchase the rights to develop it into both a stage play and later a film. *Time* magazine had given the same galley proofs a good review prior to its release (Time 1961). A few years earlier (1955), Kirk Douglas had established his own production company (Byrna Productions, named after his mother), which produced the hit, *Spartacus* (1960). There was a "bidding war" on the rights acquisition because playwright, Dale Wasserman, was also interested in developing Kesey's book into a play. Douglas made a deal with Wasserman (who also wrote the script for "Man of La Mancha") that he would hire him to write the Broadway play, and that Wasserman would retain all rights thereafter to the play, while Douglas retained the screen rights to the work (Vallance 2009). The play opened in Boston, and then opened on Broadway for a five month run. Kirk Douglas' performance was not well received because he apparently was unable to be "unlikable" (Kyselyak 1997). When Douglas moved on to try to develop the novel for the screen, he approached Czechoslovakian director, Milos Forman, and sent him the book to review. The book was apparently confiscated by Czechoslovakian customs, and Forman assumed that Douglas was not reliable; Douglas assumed Forman was rude for never contacting him about the project (Kyselyak 1997). Douglas was unable to get any major studio interested in the film.

1975: The Film

A decade later, Kirk Douglas' son, Michael, was interested in pursuing the film project as a producer, and Michael co-produced the film with Saul Zaentz. They eventually got United Artists to back the film on a small budget; inadvertently, he and Zaentz contacted director, Milos Foreman (now living in the U.S.) without knowing he had been previously contacted by Kirk Douglas. The screenplay was written by Bo Goldman (Kyselyak 1997) after Kesey declined to work on the screenplay due to arguments over the approach (Arbeiter 2016).

Twentieth Century Fox was interested in the film but only on the condition that the ending be re-written so that McMurphy lived; the producers thought that was a mistake and went with another company—United Artists. Various actors were approached for McMurphy, such as Gene Hackman and Marlon Brando (who turned it down). Foreman even considered Burt Reynolds. But Jack Nicholson won the role after the producers saw his work in *The Last Detail* (1973). Nicholson's approach to the character earned him an academy award for Best Actor (the film swept the Academy Awards that year).

As for Nurse Ratched, a relative unknown—Louise Fletcher—got the part after several well-known actresses turned it down, including Anne Bancroft, Colleen Dewhurst, and Geraldine Page. At this time, Fletcher was cast in Robert Altman's *Nashville* (1976), and withdrew from the film to play Ratched. Lily Tomlin had been contemplated for *One Flew Over the Cuckoo's Nest*, but took Fletcher's place in *Nashville* (Milos Forman 2017). Fletcher also had an unusual approach to the character and played her in a very quiet and understated manner (Kyselyak 1997), which also won an award for Fletcher of Best Supporting Actress.

The film was shot on location in Oregon State Hospital, and it cast Dr. Dean R. Brooks—the hospital's actual superintendent to play McMurphy's doctor. Brooks was very familiar with the novel, and was eager to participate. Several of the hospital's patients were cast as extras or worked as crew members (Kyselyak 1997). (This would not be feasible today due to privacy rules.)

A non-actor in the film, Mel Lambert (who met Michael Douglas on a train), a used car salesman in the area with ties to the Native American community locally, suggested Will Sampson for the part of Chief Bromden (Kyselyak 1997). Sampson was 6 foot 4, and made his debut in the film. Forman ensured that the entire cast and crew prepare for the project by first watching the 1967 documentary (Wiseman 1967) *Titicut Follies* (Milos Forman 2017), described by its distributor as a "graphic portrayal of the conditions that existed at the State Prison for the Criminally Insane at Bridgewater, Massachusetts...and documents the various ways the inmates are treated by the guards, social workers and psychiatrists" (Zipporah Films 2017).

Filming began January 4, 1975. An "Actor's Studio" approach to the film was used in which the actors prepared for the roles by living as patients in the Oregon State Hospital psychiatric ward for about ten days prior to principal filming; they lived as patients and interacted with the real patients in the hospital. Some of the actors (including Danny Devito) began having some actual mental health issues as a result of their institutionalization, and even sought help from Dr. Brooks during and after the filming (Kyselyak 1997). The hospital had a long history (Goeres-Gardner 2013a, b; Oregon State Hospital Museum 2017), and was originally called the Oregon State Hospital for the Insane (1862); it was then moved and rebuilt, opening as the Oregon State Insane Asylum (1883), and then renamed Oregon State Hospital in 1913, after there were numerous petitions by its Board of Directors to soften the name. This hospital was typical for its time, and known for practices that included ECT, lobotomy, hydrotherapy, and eugenic sterilization. Lobotomy and eugenics programs continued right up until the early 1980s, when lobotomy was abolished in 1981, and eugenics was abolished in 1983. However, it remained overcrowded well into the 1990s. The changing patient population of the hospital is reflective of the change in psychiatry patients, which is discussed more under History of Medicine. Both Nicholson and Fletcher observed ECT while at Oregon State Hospital (Hoad 2017). Forman also wanted to film unscripted group therapy sessions, which were essentially improvisation workshops for the actors to develop their characters; these sessions were filmed unbeknownst to the actors. Dr. Brooks had also helped to arrange for the actors to shadow some of the patients, so they could gain entry into the experiences of psychiatric patients in the institution. TCM's retrospective on the film states the following:

> The film was shot almost entirely on location at the Oregon State Hospital in Salem, Oregon-the very hospital where Kesey set the novel-with real patients and doctors participating as extras to add to the accuracy and the atmosphere of verisimilitude. "The realism of the location rubbed off," says Douglas. The actors participated in therapy sessions with the patients and carried their characters outside of shooting. "We got so involved that some of the actors actually took on the psychotic problems of the patients they played," according to Fletcher. Nicholson remarked that: "Usually I don't have much trouble slipping out of a film role, but here, I don't go home from a movie studio. I go home from a mental institution. And it becomes harder to create a separation between reality and make-believe. (Axmaker 2017)

Synopsis

A prisoner (R. P. McMurphy) fakes mental illness in order to get out of his prison sentence at a prison work farm. He is sent to the Oregon State Hospital for psychiatric evaluation and soon discovers the environment is more oppressive than where he came from due to the many forms of psychological control wielded by the ward nurse in charge, Nurse Ratched. McMurphy tries to disrupt routines on the ward, and befriends several patients, including Chief Bromden, a Native American male who feigns to be a deaf-mute. Although the hospital's medical director concludes McMurphy is not mentally ill and wants to send him back, Nurse Ratched argues that his sociopathic behaviors can still be helped, and by turfing him back to the prison, the hospital would abdicate its obligation to try to help him. Thus, McMurphy is admitted (committed) as an involuntary patient and subjected to questionable therapies such as ECT without consent. McMurphy decides he wants to escape and sneaks two women onto the ward through a window he bribes a guard to open. McMurphy "parties" on the ward with the women and other patients, trashing it, and in an impulsive decision, arranges for one of the women to sleep with a young virgin on the ward, stuttering Billy Bibbett under complete control by Nurse Ratched and Billy's mother. After an alcoholic-laced party, McMurphy falls asleep by the open window "waiting" for Billy to complete the act so he can leave with the woman (Candy), who is supposed to drive him to the Canadian border for his escape. He awakes the next morning to see that the "window" of opportunity has passed, and Nurse Ratched has arrived back on the ward to see the chaos. She finds Billy and Candy asleep and naked in his bed. She quickly resumes control and order, and questions Billy, making clear she will need to tell his mother. With this knowledge, Billy becomes highly agitated, begs her not to disclose his sexual act to his mother, and when Nurse Ratched insists she must tell his mother, he commits suicide. This prompts McMurphy to assault the nurse and attempt to strangle her. This results in his being lobotomized, rendering him vegetative. Chief Bromden, who is large and 6 ft., decides he's had enough and will escape by lifting an impossibly large marble bathroom fixture and tossing it through a window. Prior to his escape, he smothers McMurphy with a pillow in an act of mercy (euthanasia). The film leaves the audience shaken and haunted and questioning psychiatry as a field.

The Social Location of *One Flew Over the Cuckoo's Nest*

When Kesey is writing the novel, One Flew Over the Cuckoo's Nest, it was the early 1960s at the start of the Kennedy administration, the height of the Cold War, and when the Civil Rights movement begins to get heated. The Korean War had only just ended, and the Bay of Pigs was in the news. The novel is published the same year as the Cuban Missile Crisis; schools are doing "duck and cover" drills, and Kesey's generation—the Beat Generation (see Chap. 4), is questioning authority. As discussed earlier, Kesey's introduction to LSD from his research participation leads him into heavy recreational drug use from this study. His experiences working in a psychiatric ward in California reinforced his questioning of authority and behavioral norms, which informs the novel. In 1962, a novel about a psychiatric ward and abuse of authority over a vulnerable population resonates with a general public witnessing desegregation at Little Rock High School (1957); white supremacists beating up the first wave of Freedom Riders (1961); and a host of non-violent protests through this period that culminate into the March on Washington (August 28, 1963), just a few months after the novel is published. On November 22, 1963, what some authors point to as the actual beginning of the "Sixties" as we know it, President Kennedy is assassinated, which begins a dramatic sociological sea change in the United States, just as Kirk Douglas begins to perform in the play (debuting November 13, 1963). Between 1964 and 1966, the Baby Boomers begin to come of age, and seem to "go crazy", embracing the Hippie movement (actually spearheaded by the Beat generation). Recreational hallucinogens, prompting "trips" and new behavioral norms mirroring psychosis become part of the counter-culture movement. By the 1967 "Summer of Love", the youth culture has taken over and spreads to all corners of the country. A year for the history books is 1968: what many cite as a "tipping point" year in which it seemed as if the United States was falling into uncontrolled madness. Martin Luther King is assassinated April 4, prompting riots across the county; Robert F. Kennedy is assassinated June 6, 1968; that August (26–29) the Chicago 1968 Democratic Convention was the most violent political conventions in recent U.S. history as

Vietnam Veterans protest and are beaten up by "authority", and the Democratic Party implodes over the war. By this point, the war in Vietnam had escalated, as the "madness" of the conflict is shown on television each night. To adults living in 1968, it seemed as if the "lunatics" were running the asylum. Interest in <u>One Flew Over the Cuckoo's Nest</u> only magnified as everything that seemed "normal" a decade ago now feels "crazy" and everything that seemed "crazy" a decade ago is now considered "normal"—particularly attitudes about sexuality, women and minorities. In fact, the social norms had not been so dramatically changed in so short a timeframe as they were during this period. Films of 1968–69 offer a glimpse of the social psyche: *Easy Rider* (1968), also starring Jack Nicholson, about questioning traditional authority; *Planet of the Apes* (1968), reflecting cultural anxieties over Civil Rights; *2001: A Space Odyssey* (1968), involving themes of evolution, autonomy of machines versus man, and long interludes of psychedelic imagery that mirrored an LSD "trip"; *Night of the Living Dead* (1968), a horror film about society going mad; *Rosemary's Baby* (1968), a prenatal horror film about a pregnant woman losing control of her body; *Midnight Cowboy* (1969), a disturbing cult classic about social misfits.

And then, just as it looked as though the madness was unstoppable, Richard Nixon, a stable figure from the 1950s, who ran as a "law and order" candidate, won the 1968 U.S. Presidential election in which the "silent majority"—adults who felt no one was in control of the asylum—voted for an authority figure to take control of the country. By 1969, in the early days of the Nixon administration, two August events reinforced the themes of "madness" again: The Manson Murders (August 9–10, 1969), in which bizarre home invasions and murders by members of a hippie "cult" became major news when celebrity, Sharon Tate and several of her celebrity friends were victims, including a middle aged couple the next night in a Los Angeles suburb. The Manson murders and later trial held all in thrall (as in the "O. J. Simpson trial" in the 1990s) with respect to dominant media coverage; and then a week later, Woodstock (August 15–18, 1969) would become the closing act of the 1960s. Ultimately, living through the late 1960s was akin to navigating through madness—a generation that was grappling with existential questions and "acting out".

Early 1970s: Authority in the Watergate Era

Between 1970 and 1975, the timeframe in which the film is being developed and shot, questioning, and dismantling authority occurred during the Watergate era, in which a President faced impeachment for obstruction of justice after a bungled burglary at the Watergate building, on June 17, 1972 turned into an involved cover-up. This period of history comprised what was called a "constitutional crisis" in which the President of the United States became embattled against the two other branches of government. By July 1974, The House Judiciary Committee had adopted three articles of impeachment against President Nixon: obstruction of justice, abuse of presidential powers, and hindrance of the impeachment process.

The film project for *One Flew Over the Cuckoo's Nest* is being done at the height of the Watergate drama, and is released November 19, 1975, less than two years after Nixon resigned on August 8, 1974, during the Ford administration. The country had now been through what Ford called a "long national nightmare," and a free press and journalism ethics acted as the ultimate "check" on uncontrolled power, ultimately toppling the Nixon Presidency. By 1975, the audience was eager to see a film whose main character challenged authority; it's important to note, however, that though released in 1975, *the film is faithful to the timeline of the book, and takes place in 1963*. Footage with respect to civil rights issues and radio replays of that year's World Series dots the film.

For a 1975 audience, Civil Rights had now spread to other vulnerable populations advocating for rights, and this included the Native American population (see further), and even the prison population whose civil rights demands became known during the Attica prison riot (September 9–13, 1971). The same year *One Flew Over the Cuckoo's Nest* premiered, the book A Time to Die (Wicker 1975), was published, which chronicled this infamous prison riot, initiated to demand better living conditions, including better quality food, sanitation, and medical treatment, for the incarcerated. As discussed under History of Medicine, prisons would become filled with untreated mental health patients and began to function as makeshift psychiatric hospitals in the wake of de-institutionalization and lack of access to mental health services. Attica was fresh to this audience, and also referenced in *Dog Day Afternoon* (1975), in which an ex-convict robbing a bank (played by Al Pacino) takes hostages chanting "Attica, Attica, Attica."

The Native American Context

You can't discuss this film without discussing the importance of the Native American context, particularly when teaching this to healthcare trainees who will encounter health disparities in indigenous populations. Kesey's choice to make a Native American (a.k.a. American Indian) the narrator in the novel and a major character was ahead of its time, as Native Americans were not a group in the early 1960s that were visible as an oppressed minority (TES 2017; Digital History 2016; The Zinn Education Project 2017). However, Kesey's choice to call Bromden "Chief" was also a reflection of the times. Cultural competence when teaching this film demands recognition that calling Native American males "Chief" is a derogatory term that is insulting; McMurphy in the film also denigrates Bromden further by raising his hand and saying "How" and making faux "Indian war whoops" when he encounters Bromden for the first time. It is also worth unpacking the terminology of Native American and American Indian; some tribes have a preference for the latter because it makes clear that they are *Americans*. For the purposes of this chapter, I will use the term Native American unless referencing historic text or agencies that used "Indian". Of note, the term "Indian" to refer to indigenous peoples in the United States is reportedly linked to Christopher Columbus, who recognized differences in appearance between Europeans and

Native Americans, and wondered in his geographic bubble if they were from "India"—and began calling them "Indians" in his famous letter to Luis De Sant, announcing his discovery (Columbus 1493).

When Kesey is writing the novel, conditions on Native American reservations were deplorable; many reservations had no plumbing, poor sanitation and no healthcare access. Under the Eisenhower Administration, new policies were developed to remove Native Americans off reservations and into the urban centers to encourage assimilation, which was a failure; by 1961, under the Kennedy Administration, the policy was discontinued, and the United States Commission on Civil Rights noted that for "Indians", "poverty and deprivation are common." (University of Groningen 2012).

There were stirrings of activism when the novel was published that may have been noticed by Kesey, but weren't attracting much media attention. In 1959, the Tuscarora tribe (upstate New York) lobbied against a reservoir project on reservation land; in 1961 a National Indian Youth Council introduced the phrase "Red Power" and copying African American civil rights demonstrations, staged marches and "fish-ins" to protest efforts to abolish Native American fishing rights (Digital History 2016).

Native American activism became far more visible years later, in 1969, when a group had occupied Alcatraz Island for 19 months in protest. They were demanding the "return of Alcatraz to the American Indians and sufficient funding to build, maintain, and operate an Indian cultural complex and a university." The activists, who called themselves Indians of All Tribes, offered to buy Alcatraz from the federal government for "$24 in glass beads and red cloth" (The Zinn Education Project 2017).

The American Indian Movement (AIM) thus began, which focused on the mistreatment of Native Americans by the U.S. government through staged protests. In 1970 at Mount Rushmore and at Plymouth Rock on Thanksgiving Day, which was established as a "Day of Mourning". In 1972 the Trail of Broken Treaties Caravan staged a six-day occupation of the offices of the Bureau of Indian Affairs (BIA) in Washington, D.C.

During this period, Native American issues permeated more into mainstream media and film. *Little Big Man* (1970) starring Dustin Hoffman was a major Hollywood film about American abuses of the Native American population. Though historically flawed, the film was widely acclaimed. Around this timeframe, a series of "Make America Beautiful" Public Service Announcements aired on television, featuring the "American Indian" and "his proud heritage" decrying litter and pollution (PSA 1971; Waldman 1999). Images of the tearful Native American male in traditional tribal costume amidst garbage and polluted environments became part of the public consciousness. "Make America Beautiful" was a campaign started by First Lady, "Ladybird" Johnson as part of her American beautification project, one of the first to draw attention to environmental problems.

In 1973, 250 Sioux Indians returned to Wounded Knee and led an occupation of that territory in honor of the 1890 massacre of hundreds of Lakota men, women and children. Another event that same year occurred at the Academy Awards ceremony (televised March 27, 1973). Marlon Brando, who had won Best Actor for *The Godfather* (1972) refused his award by sending Native American activist, Sacheen Littlefeather up to the podium to refuse on his behalf (Thomas 2016; Arbuckle 2017). Brando had published his Oscar refusal speech in the *New York Times* three days later (Brando 1973). Littlefeather, in front of 85 million viewers, said that Brando "very regretfully cannot accept this generous award, the reasons for this being...are the treatment of American Indians today by the film industry and on television in movie reruns, and also with recent happenings at Wounded Knee" (Bort 2016).

Brando's full speech in the *New York Times* started as follows:

> For 200 years we have said to the Indian people who are fighting for their land, their life, their families and their right to be free: Lay down your arms...When they laid down their arms, we murdered them. We lied to them. We cheated them out of their lands. We starved them into signing fraudulent agreements that we called treaties which we never kept. We turned them into beggars on a continent that gave life for as long as life can remember. And by any interpretation of history, however twisted, we did not do right. We were not lawful nor were we just in what we did....

Another key excerpt:

> What kind of moral schizophrenia is it that allows us to shout at the top of our national voice for all the world to hear that we live up to our commitment when every page of history and when all the thirsty, starving, humiliating days and nights of the last 100 years in the lives of the American Indian contradict that voice?

Brando was heavily criticized for using the Academy Awards as his political stage, but his ploy worked: it drew attention to the Native American plight. And by the time *One Flew Over the Cuckoo's Nest* became a film, Chief Bromden's character resonated even more with the film's audiences, who had far more awareness of Native American issues than when the novel was published over a decade earlier.

Throughout the 1970s, major legislation and court cases acknowledge moral obligations to Native Americans. In 1972, the *Indian Education Act* is passed; by 1976, the *Indian Healthcare Act* is passed, which recognized health disparities that persist today; in 1978, the *Indian Child Welfare Act* is passed; and a series of court decisions held that Native Americans had the right to sovereignty and tribal self-government.

In the novel, Chief Bromden's father, described as "Chief Tee Ah Millatoona," had married a white woman and took her surname, Bromden; he is not a Chief himself, and is assigned the name by the other patients. Accordingly (Native American Indian Association 2017):

> Traditional use of the title 'Chief' is an honor restricted to those leaders of Native American tribes who have received the title through tribal selection or inheritance...use of the title 'Chief' by persons who are not tribal business or traditional leaders is considered offensive. Non-Indians addressing individuals who are not federally- or state-recognized tribal leaders or elders by the title 'Chief' is also considered offensive to traditional Native American people.

Thus, the character Chief Bromden is so-named as a product of the times, not unlike derogatory, and now-antiquated terms for African Americans in other works of their period. Will Sampson, who died in 1987 (Knickmeyer 1987) who plays Bromden, was a Creek Indian who plays the character masterfully. Chief Bromden, as written and performed, is a symbol of Native Americans in this timeframe: No one seems to be hearing them, and they feel voiceless. Finally, it would be important to discuss Chief Bromden's character within the current context of Native American activism such as the recent protests surrounding the Dakota Pipeline Access that took place from April, 2016–February 2017.

History of Medicine

There are four History of Medicine contexts to cover when teaching this film: History of psychiatric patient placement; psychiatric treatment; and health disparities in psychiatry, such as the treatment of prisoners. There is an entangled theme of institutionalization and imprisonment that should be unpacked for discussion. It's also important to highlight that within medicine, psychiatry establishes itself as the second subspecialty after surgery, but that "asylums" for the social misfit and/or mentally ill predate the subspecialty of psychiatry, which were generally run by the state.

History of Psychiatry and Placement

When reviewing this history of medicine context, it's necessary to first discuss placement/housing of the mentally ill as the common thread that drove treatments and legislation (Knapp et al. 2011; Lamb and Weinberger 2005; Martinez-Leal et al. 2011; Novella 2010; Torrey 1997; Interlandi 2012; Shatkin 2013). Confinement versus community living was always a difficult question. Until about 1600, the mentally ill lived in the community unless they were thought to be dangerous. They were seen as a family's burden, and not society's. Around the 1600s, Europeans began to isolate the mentally ill by chaining and confining them as though they were prisoners. This became the norm, and by the 1700s, institutionalization was the "standard of care". By the late 1700s, early ethical questions arose surrounding what some felt were inhumane treatments. French physician, Phillippe Pinel, frees patients from chains, and he changes the environment to sunny rooms, allowing patients to exercise and roam around. Until the early 1800s, asylums flourish in the U.S. and are intended for wealthy families; almshouses wind up as makeshift institutions for everyone else that society doesn't accept. By the 1820s, public asylums are established, and in the 1840s, social welfare advocate, Dorothea Dix helps to establish humane treatment and 32 state hospitals in the U.S. In 1844, the Association of Medical Superintendents of American Institutions for the Insane (AMSAII) founds the *American Journal of Insanity,* which begins to promote research and best practices. This organization later becomes the American Psychiatric Association. By the late nineteenth century, responsibility for the mentally ill fell under jurisdiction to state asylums, which are essentially repositories for the old and those with tertiary syphilis. In 1879, the first psychiatric hospital in the United States was established, which was Belvue in New York City.

By the 1920s and 1930s, questions surrounding home care arise because of terrible overcrowding in these institutions, and by the 1950s, there is a push for community mental health centers as a dominant model; this was an international movement afoot to reform the "asylum-based" model to a community-based model. It became clear by this point that many psychiatric patients could live in the community and have a greater sense of belonging rather than be hidden or isolated in an institution. Community based care did not take hold in the United States until the mid-1960s, after passage of the *Community Mental Health Centers Act* of 1963, signed into law by John F. Kennedy, who was inspired to see it pass due to the plight of his sister, Rosemary Kennedy, who had been mentally disabled (see further) but lobotomized when she was 23 in 1941 (Larson 2015). Thus, *One Flew Over the Cuckoo's Nest* represents a picture of mental health care in the United States just prior to de-institutionalization. Under the 1963 law, closure of state psychiatric hospitals in the United States were codified, and strict standards were passed so that only individuals "who posed an imminent danger to themselves or someone else" could be committed to state psychiatric hospitals. Thus enters the era of much stricter criteria for involuntary hospitalization. However, based on these criteria, even competent and capacitated patients could still be hospitalized against their will if they meet this standard (see under Clinical Ethics section). This Kennedy Administration law coincided with the sweeping social reform legislation under the Johnson Administration, and the passage of Medicaid (1965), which offered better reimbursement for nursing homes than for mental hospitals. Between 1950 and 1980, there was a sharp decline in mental health patients residing in institutions, compared to other types of facilities such as community mental health centers; and smaller supervised residential homes by community-based psychiatric teams. By 2000, there were only 22 beds per 100,000 persons in psychiatric hospitals; in 1955, there were 339 beds per 100,000 persons.

In 1972, Social Security Disability Insurance (SSDI) expanded to include the mentally disabled, and the *Social Security Act* was amended to provide coverage for people who didn't qualify for benefits: both led to community living for the mentally disabled. Moffic (2014) contextualizes *One Flew Over the Cuckoo's Nest* this way:

> The film wasn't made until 1975... this was the heyday of community mental health—there were hundreds of federally funded centers across the US, well on their way to providing comprehensive services in the communities instead of in the state hospitals. By 1981, President Reagan, began the dismantling of these centers by substituting block grants, whereby states could use these funds for other public services, such as highways. Around 1988, two different societal trends affecting psychiatry gained traction in the US: the recovery movement and managed care, emphasizing consumer empowerment and management control of treatment, respectively.

Funding notwithstanding, another major factor in the closing of asylums and institutions had to do with redefining what constituted mental illness to begin with, and the recognition that "social misfits" were often healthy individuals rebelling against systemic and societal oppression of sexuality and gender preference; women; different cultures and races; and so forth. In significant ways, social reform movements such as civil rights, including feminism and gay rights; and then later, recognition of the autism spectrum and other learning disabilities, "cured" mental illness because behavioral norms were expanded and diversified. Combined with legislation demanding patients be treated in the "least restrictive setting" in the community, as well as better treatments, including pharmacological interventions, those patients requiring hospitalization significantly shrunk.

Treatment Issues

One Flew Over the Cuckoo's Nest is also a study of the history of psychiatric therapies and standards of care, which have significantly evolved. Treatments to focus on when teaching this film would be ECT, which became popular around the 1940s, and which proved to be beneficial once proper candidacy was established with informed consent. In the film, we see ECT being performed and are horrified; unfortunately, it is reflective of the timeframe, but not how ECT is currently performed. ECT has been shown to be beneficial, and is done using completely different methods with informed consent. Many psychiatrists believe this film served to repel patients against ECT, who some believe could have benefited (Hawksley 2014; Swaine 2011).

The same is not true for lobotomy, which was performed on R. P. McMurphy—one presumes—as a "behavioral surgery" to control him. Lobotomy is an American medical embarrassment, which was pioneered in the United States by Walter Freeman, and which ultimately proved to be typically disastrous (medically and ethically) for patients and both the psychiatric and surgical profession as a whole. Lobotomy was a surgical procedure practiced from the 1940s–70s as a purported "treatment" for behavioral control on a highly heterogeneous, vulnerable population without any established benefit, defined candidates, and no informed consent. The initial motivation for lobotomy was to reduce crowding in asylums by using a surgery to "cure" troubling behaviors in patients so they can be returned home. The adage: "lobotomy gets them home" was the sales pitch (Goodman 2008). Troubling patient behaviors could range from "feminism" to genuine psychiatric symptoms, or from low IQ and intellectual impairment to autistic spectrum disorders. If anyone is a candidate, no one is a candidate. For example, one of the most famous lobotomized patients was Rosemary Kennedy who underwent lobotomy in 1941 (when she was 23) without consent or assent; her father (Joseph Kennedy) personally asked Freeman to perform the surgery, which left her far more debilitated to the point where she was hidden away in a Catholic nursing home for the rest of her life (Larson 2015). Even in 1941, lobotomy carried questionable benefit; definite harms, and there was no established patient candidacy for the procedure. The history of lobotomy is one of the most important examples of medical abuse and violation of the Principle of Nonmaleficence (the obligation not to knowingly cause harm), which I explore more in Healthcare Ethics on Film. At the same time, it is a relatively unknown story for most medical students, and is seldom covered in either neurosurgical curricula or even most psychiatry curricula. This is because it is relegated to a "history of medicine" topic. For these reasons, when teaching this film, it would be important to assign the PBS documentary, *The Lobotomist* (2008), widely accessible online, alongside this film, or as a supplemental viewing. Key factors to point out to students surrounding the history of lobotomy are these: (a) it was an innovative therapy in the absence of any other treatment; (b) conditions in asylums were abysmal when it was initially pioneered, and reducing patient populations in these places was thought to be a benefit; (c) published results were skewed to downplay negative results in a timeframe where peer review was not optimal.

For a psychiatric trainee audience, *One Flew Over the Cuckoo's Nest* also demonstrates "Milieu Therapy," developed in the 1950s, which advocated for a supportive asylum environment. Finally, the film also demonstrates how medications (probably thorazine) were used for behavioral control (Shatkin 2013), and the potential for psychotherapy to be misused—particularly with the character Billy Bibbet, who ultimately commits suicide when Nurse Rached threatens to "inform his mother" about his sexual tryst. To this day, questions surrounding biopsychiatry and drugs used for chemical restraint remain in debate.

With respect to the history of ECT, it's important to discuss current usage and its benefits with the right population of patients. Assigning current review articles, as well as having a psychiatrist faculty member present to discuss and explain the procedure as it's been perfected over the years, would be an important teaching enhancement for a medical student audience, in particular. According to Moffic (2014):

> The traumatic depiction of ECT in the movie, used as much for control as treatment, had a major impact on the subsequent lack of availability of ECT, despite the development of much safer delivery systems and its potential for lifesaving quick benefits. Many feel that the development and use of other related treatments, such as transmagnetic stimulation, vagal nerve stimulation, and surgical deep brain stimulation, have been slowed by their association with ECT. Meanwhile, we continue to learn more about the long-term adverse effects of various medications, and many psychotherapies are becoming less available.

Ultimately, righting of historic abuses of psychiatric placement and treatments lay in legislation that limited involuntary hospitalization as well as codified ethical guidelines surrounding informed consent, competency and capacity assessment.

Health Disparities in Psychiatry

This film wittingly, and unwittingly, demonstrates health disparities in psychiatric treatment in three populations: the prison population, of which McMurphy is a representative; the Native American population, of which Chief Bromden is a representative; and the absence of any African American patient in the film. (It would be expected that on a male ward, there would be no female patients, so a discussion of women in psychiatry would be better with a different psychiatry film).

When this film takes place (1963), access to psychiatric care in the prison population was based on an asylum model in which the "criminally insane" were separated from the rest of the prison population and placed in separate institutions. In McMurphy's case, he is being sent from prison to Oregon State Hospital for an evaluation as to whether he should be admitted. As is often the case, this prisoner is malingering, hoping to be permanently assigned to a hospital for the remainder of his sentence because he believes it will be a better environment. In his initial interview with the medical director, he is told frankly that he does not exhibit any behaviors suggestive of mental illness, but will be observed for four weeks. In a multi-team meeting, a few weeks later, after McMurphy proves disruptive and noncompliant, there is consensus that he is not "mentally ill" (and therefore competent and fully capacitated), and the physician team agrees he should be sent back to the prison. Nurse Rached disagrees, and makes an argument that he is still sociopathic, and thereby an appropriate admission. She argues that sending him back to prison may be a form of "turfing" the patient instead of trying to improve his behaviors through hospitalization. He is thus involuntarily committed, based on his sociopathic behaviors. The nurse proves correct: ultimately, his behaviors *are* moderated, alas, through lobotomy, which results in incompetence.

Today, the opposite occurs: prisons have become makeshift psychiatric wards (Navasky and O'Connor 2005) for people who are truly mentally ill, and whose behaviors get them arrested and incarcerated due to de-institutionalization, combined with lack of access to care. However, if prisoners are sent for hospital evaluation or the Federal Medical Center, they have rights (see further on). The prison population is disproportionately African American, a population traditionally barred from access to psychiatric treatment even though their mental health was dramatically affected by unfair social arrangements that perpetuated poverty and unequal treatment across the societal spectrum. This trend was only beginning when the film was released in 1975, and has only become far more pronounced.

With respect to disparities in the Native American population, Chief Bromden relays his father's struggles with alcoholism, which continues to afflict this population; Native Americans die of alcohol-related causes roughly six times more than the national average while rates of suicide and depression in this population remain much higher due to continuing issues with disenfranchisement (Moffic 2014). (See earlier under Social Location).

Clinical Ethics Issues

There are numerous clinical ethics issues to untangle in this film, that include competency and capacity; nursing ethics; beneficent versus maleficent therapies, and the ethical treatment of prisoners. Other discussions may surround suicide and euthanasia. As this chapter's title suggests, if you needed to limit discussion due to time constraints, I would suggest focusing on the assessment of competency and decision-making capacity. In this film, a competent prisoner patient is involuntarily committed to a psychiatric state hospital, and ultimately is provided with non-beneficial treatments for the sole purpose of behavioral restraint (ECT and later lobotomy). Arguably, the only truly ethically defensible action is Chief Bromden's "mercy killing" (euthanasia) in the end, given that McMurphy's personhood is gone.

Competency Versus Capacity

Competency is a characteristic or property one possesses, while capacity is best understood as an "ability" that can be limited due to a number psychosocial or physiologic barriers. There are many individuals who are not legally competent, such as those who are severely intellectually or neurologically impaired, ranging from congenital injuries to acquired brain injuries resulting in persistent vegetative states. In these cases, competency cannot be restored, and a surrogate decision maker is needed. In many such cases, a court-appointed guardian is assigned. Legally, all emancipated adults are presumed competent unless proven otherwise. Thus, any adult under 18 is considered not competent unless s/he is an emancipated minor. Finally, anyone who is arrested and awaiting trial must be determined to be competent to stand trial as part of his/her constitutional rights. Thus, competency hearings in the justice system occur frequently. All competent adults are presumed to have decision-making capacity whereby they must meet the U-ARE criteria (see Table 5.1) with respect to individual decisions; barriers to capacity can include literacy, numeracy, and underlying medical conditions—including psychiatric—that interfere with clear thinking. Generally, competency hearings in the courts require expert opinions on whether individuals are competent, which also relies on meeting U-ARE standards (Roth et al. 1977). It is important to emphasize that the courts typically do *not* consider mental illness or addictions per se to be grounds for incompetence. Thus, many patients who are mentally ill are still considered competent, and presumed to have capacity unless proven otherwise; this includes individuals who make terrible decisions who are a danger to themselves or others—the threshold for involuntary holds or hospitalization. It's critical in teaching to explain that competency and involuntary commitment are mutually exclusive: one does not depend on the other, although they frequently occur together. Competency and sociopathic behaviors can also co-exist, which is an important discussion.

The critical question that the film addresses is whether R. P. McMurphy is competent. In the film, there is a consensus expert opinion that he is indeed competent, but, as his nurse argues, his sociopathic behaviors still qualify him for involuntary commitment, which she argues, would be a societal benefit and a goal of care. When McMurphy learns he has been committed, he is shocked to discover that most of the other patients on the ward are *voluntary* patients. They are in the hospital because they deem themselves unable to function, but are still legally competent, even though some prove later, to be a "danger to themselves" as Billy Bibbett proves to be when he commits suicide.

Every clinical ethicist, at some point, must make clear during the process of clinical ethics consultation involving capacity assessment or behaviorally challenging patients, that competency and "capacity" cannot be terms used interchangeably, and are not the same thing. Yet these ideas are typically conflated. In the healthcare setting, many healthcare providers have moral distress when they are legally constrained from hospitalizing patients they feel could benefit. All competent patients are legally and ethically permitted to leave Against Medical Advice. Patients must meet strict criteria and conditions for a "24-/48-/72-h hold". Indigent patients seeking shelter will frequently present to emergency rooms, in fact, and claim they are "suicidal" or intend to harm the public in some way for the sole purpose of being hospitalized for shelter. Thus, malingering to get hospitalized, as R. P. McMurphy has done, is nothing new. At the same time, even patients with serious addictions and untreated schizophrenia may not meet grounds for involuntary hospitalization. Of course, previously competent patients can become incompetent if they permanently lose consciousness, but determining incompetence is a fairly high bar.

All competent adults are presumed to have decision-making capacity unless they fail to meet the U-ARE criteria, which may occur when there are barriers that may include language or literacy. Decision-making capacity is task-specific and may be an ability some incompetent patients still have (as in a 17 year-old patient) depending on the decision involved. There is also no single standard for capacity, but the more consequential the decision, the higher the standard we would demand. Competent patients may lose decision-making capacity due to barriers, which can include underlying mental illnesses that could be treated to restore decision-making. The nuances of decision-making capacity are discussed more in Chap. 7, which deals with dementia and gradual loss of autonomy and capacity.

Finally, it's important to note who determines competence: experts solicited by the courts or justice system determine competency. All other assessments in a hospital setting surround *decision-making capacity*, which is a medical determination that rests with the Attending physician of record who may elect to request input from other consultants, including psychiatry. When consulted, the role of psychiatry is to determine whether there are any underlying mental health problems that are *barriers to capacity* to help the Attending physician in his/her assessment. When capacity is in question, psychiatry provides an expert assessment to see if a particular decision is based substantially on delusions or depression, or other psychiatric conditions that could be treated. In McMurphy's case, he is sent for observation to be evaluated for evidence of mental illness to help the state determine whether he should be hospitalized or returned to prison; the hospital sees no evidence of mental illness. But as a state hospital, McMurphy is in their charge, and the hospital has the authority to admit McMurphy because of his sociopathic behaviors. It begs the question as to how "voluntary" the other patients' status is; if they are non-compliant, would they, too, be committed against their will?

Nursing Ethics

This film demonstrates extreme patient violence against nurses, which is a real problem in behavioral health patients. When such patients are in a hospital setting, and they are a threat, chemical restraint as a *behavioral intervention* is still ethically permissible, and is why McMurphy was lobotomized; it was done as a behavioral intervention. It's worth asking, when teaching this film: Is Nurse Ratched evil? Louise Fletcher, the actress who plays her, suggests in interviews that Nurse Ratched may be an "instrument of evil" as part of the medical system, but that she believes she is doing good. The viewer may be torn about whether Nurse Ratched deserves to be strangled by McMurphy, who sees her as the direct cause of Billy's suicide. In my own experiences, some of the most wrenching clinical ethics cases involve violent behavioral health patients who have assaulted nurses, and an ethics opinion about restraint is requested. Frequently, what is ethically permissible may not be in the patient's best medical interests at all, but may be necessary for overall patient safety in a hospital setting when weighing competing ethical principles. Given that lobotomy was a "standard of care" at the time for behavioral control, was it ethically permissible as a behavioral intervention therapy after attempted murder of a nurse? Today, we still use chemical restraints and physical restraints to modify violent behaviors.

Beneficence

This film raises questions about what treatments maximize benefits and minimize harms in a psychiatric hospital; the same questions can still be raised about many psychiatric therapies, particularly when balancing risks of side-effects of pharmaceutical therapies used. The exploration of beneficence in the next section of films offers far deeper discussions about what "beneficence" actually means and the distinction between intentional and unintentional harms is explored. Generally, none of the therapies demonstrated in this film can be said to be truly "beneficent"—including group therapy sessions led by Nurse Ratched. Ultimately, the film reveals that all of the therapies are used as behavioral controls that may not be good for patients' overall health and wellbeing. Director, Milos Forman, who emigrated from communist Czechoslovakia to the United States prior to making the film, commented that the setting of the film and mindset of the characters was familiar to him because he grew up in a dictatorship and understood what behavioral controls were about.

Ethical Treatment of Prisoners

As discussed above, when this film debuted, the Attica prison riots were fresh, in which rioters demanded better medical treatment, including mental health treatment. Another event that same year (1971) was the Stanford Prison Study, in which college students volunteered to play the roles of either prison guards or prisoners (Haney et al. 1973). The study was halted early because the students playing the "guards" began abusing the "prisoners" to the point where harm was being done (Haney et al. 1973). This study, years later, explained some of the events that occurred in Abu Ghraib (2005) when young American soldiers tortured Iraqis who were rounded up and held in that prison compound without due process during the Iraq War. There is also a long history of prisoner patients being exploited for medical research, and by 1975, when the film came out, the National Commission for the Protection of Human Subjects had just formed, which ultimately defined vulnerable populations, which includes prisoners.

Ethical treatment of prisoner patients requires that they are treated the same as any other patient; this means they are permitted to participate in research studies so long as they have informed consent, and the study is IRB approved; I discuss the history of research ethics and the formation of the Belmont Report guidelines in Chap. 10 (about the film, *Awakenings*). Prisoners do have considerable restrictions on their autonomy, despite having medical decision-making capacity. In the film, R. P. McMurphy as a prisoner is a ward of the state and loses his rights to decide what he does, and where he stays. However, prisoners still cannot be involuntarily treated in a hospital unless they are considered incapacitated; in which case, a surrogate decision maker would need to make decisions based on known patient preferences or what is in the patient's best medical interests. Prisoners can decide to withdraw life support, for example, even if withdrawal will "shorten their sentence" (as some wardens have been known to argue). Prisoner patients typically have guards in the room, and it is ethically permissible, for confidentiality purposes, to either request the guards leave the room at certain times, or if they cannot due to a prisoner's status (violent and/or dangerous), discuss confidential information in the presence of guards. Although they are wards of the state, prisoner patients' family members still act as surrogate decision-makers unless they are unrepresented. Even then, prisoners would require a state guardian be appointed who does not have a conflict of interest. Thus, wardens should not be in charge of medical decisions for prisoners, although many will insist that they are. In such cases, hospital attorneys may need to be involved. Wardens, however, do need to approve payment for procedures, as they are the third party payer, similar to an insurer.

Conclusions

This classic film about psychiatry ethics and competency is ideal for both a medical student and nursing student audience or any other mental health trainee (social work, counseling and psychology, etc.) which can be enhanced with a guest in class who is either a psychiatric nurse or a faculty member in psychiatry to help field questions. As noted by the *Psychiatric Times*: "More than 50 years since One Flew Over the Cuckoo's Nest was published and almost 40 years since the movie was released, the issues seem as relevant today as they were back then." (Moffic 2014.)

Film Stats and Trivia

- This film was the first to win all five of the major academy awards since 1934: Best Picture, Best Director, Best Actor, Best Actress, and Best Screenplay.
- The film had one of the longest theatrical runs in history, showing in Sweden at theaters until 1987.
- Kesey refused to see the film because he didn't agree with the screenplay; he wanted Chief Bromden to continue to be the narrator in the screenplay. In 2007 the American Film Institute ranked *One Flew Over the Cuckoo's Nest* as the 33rd Greatest Movie of All Time.
- Sampson died early from kidney failure following a failed transplant.

Theatrical Poster

Producer: Michael Douglas, Saul Zaentz
Director: Milos Forman
Screenplay: Lawrence Hauben, Bo Goldman (screenplay); Ken Kesey (novel); Dale Wasserman (play)
Cinematography: Haskell Wexler; Bill Butler (uncredited)
Art Direction: Edwin O'Donovan
Music: Jack Nitzsche
Film Editing: Sheldon Kahn, Lynzee Klingman
Cast: Jack Nicholson (R. P. McMurphy), Michael Berryman (Ellis), Peter Brocco (Col. Matterson), Dean R. Brooks (Dr. Spivey), Alonzo Brown (Miller), Scatman Crothers (Turkle), Mwako Cumbuka (Warren), Danny DeVito (Martini), William Duell (Jim Sefelt), Josip Elic (Bancini), Lan Fendors (Nurse Itsu), Louise Fletcher (Nurse Ratched).
Runtime: 133 m.

References

Arbeiter, Michael. 2016. 15 things you might not know about One Flew Over the Cuckoo's Nest. *Mental Floss*, November 19. http://mentalfloss.com/article/63639/15-things-you-might-not-know-about-one-flew-over-cuckoos-nest.

Arbuckle, Alex Q. 2016. March 27, 1973: Sacheen Littlefeather at the Oscars. *Mashable*, February 26. http://mashable.com/2016/02/26/sacheen-littlefeather-oscars/#lx3Ct3ZKSiqG.

Axmaker, Sean. 2018. One Flew Over the Cuckoo's Nest. TCM.com. http://www.tcm.com/this-month/article/296733%7C0/One-Flew-Over-the-Cuckoo-s-Nest.html. Accessed 23 Feb 2018.

Biography.com. 2016. Ken Kesey. http://www.biography.com/people/ken-kesey-9363911. Accessed 23 Feb 2018.

Books: Life in the looney bin. 1962. *Time*, February 16. http://content.time.com/time/magazine/article/0,9171,829087,00.html.

Bort, Ryan. 2016. The time Marlon Brando boycotted Oscars to protest Hollywood's treatment of Native Americans. *Newsweek*, January 23. http://www.newsweek.com/marlon-brando-boycotted-oscars-native-americans-418545.

Brando, Marlon. 1973. That Unfinished Oscar Speech. *New York Times*, March 30. http://www.nytimes.com/packages/html/movies/bestpictures/godfather-ar3.html.

Columbus, Christopher. 1493. *Letter to Juis De Sant.* http://www.ushistory.org/documents/columbus.htm.

Digital History. 2016. The Native American Power Movement. http://www.digitalhistory.uh.edu/disp_textbook.cfm?smtid=2&psid=3348. Accessed 24 Feb 2018.

Goeres-Gardner, Diane L. 2013a. *Oregon Asylum*. Images of America (series). Mt. Pleasant, SC: Arcadia Publishing.

Goeres-Gardner, Diane L. 2013b. *Inside Oregon State Hospital: A History of Tragedy and Triumph.* Mt. Pleasant, SC: The History Press.

Goodman, Barak. 2008. *The Lobotomist.* American Experience, PBS, released January 21.

Grossman, Lev. 2010. All time 100 Novels. *Time*, January 8. http://entertainment.time.com/2005/10/16/all-time-100-novels/slide/one-flew-over-the-cuckoos-nest-1962-by-ken-kesey/.

Haney, Craig, Curtis Banks, and Phillip Zimbardo. 1973. A study of prisoners and guards in a simulated prison. Naval Research Reviews, September 1–17. http://www.zimbardo.com/downloads/1973%20A%20Study%20of%20Prisoners%20and%20Guards,%20Naval%20Research%20Reviews.pdf.

Hawksley, Rupert. 2014. One Flew Over the Cuckoo's Nest: 10 things you didn't know about the film. *The Telegraph*, February 28. http://www.telegraph.co.uk/culture/film/10665661/One-Flew-Over-the-Cuckoos-Nest-10-things-you-didnt-know-about-the-film.html.

Higgins, Chris. 2012. Stop Saying 'Drink the Koolaid'. *The Atlantic*, November 8. https://www.theatlantic.com/health/archive/2012/11/stop-saying-drink-the-kool-aid/264957/.

Hoad, Phil. 2017. Michael Douglas: How we made One Flew Over The Cuckoo's Nest. *The Guardian*, April 11. https://www.theguardian.com/film/2017/apr/11/michael-douglas-and-louise-fletcher-how-we-made-one-flew-over-the-cuckoos-nest-interview.

Interlandi, Jeneen. 2012. A madman in our midst. *New York Times*, June 24. http://query.nytimes.com/gst/fullpage.html?res=9402E0DE1138F937A15755C0A9649D8B63&pagewanted=all.

Keep America Beautiful. 1971. Public Service Announcement, April 22. https://www.youtube.com/watch?v=8Suu84khNGY. Accessed 24 Feb 2018.

Kesey, Ken. 1962. *One Flew Over The Cuckoo's Nest*. New York: Viking Press.

Knapp, M., J. Beecham, D. McDaid, T. Matosevic, and M. Smith. 2011. The economic consequences of deinstitutionalisation of mental health services: Lessons from a systematic review of European experience. *Health and Social Care in the Community* 19: 113–125. https://www.ncbi.nlm.nih.gov/pubmed/21143545.

Knickmeyer, Ellen. 1987. Indian Actor Will Sampson Dies. *The Oklahoman*, June 4. http://newsok.com/article/2188096.

Kyselyak, Charles. 1997. Completely Cuckoo. Quest Productions, Warner Home Video, released December 16 (An abridged, re-cut version was released in 2002 as *The Making of 'One Flew Over the Cuckoo's Nest'* as a Special Feature on the DVD version of the film by The Saul Zaentz Film Center. Company.).

Lamb, H.Richard, and Linda E. Weinberger. 2005. The shift of psychiatric inpatient care from hospitals to jails and prisons. *Journal of the American Academy of Psychiatry the Law* 33: 529–534.

Larson, Kate Clifford. 2015. *Rosemary: The Hidden Kennedy Daughter.* New York: Mariner Books and Houghton Miflin Harcourt.

Library of Congress. 1994. Librarian announces national film registry selections. *Library of Congress Information Bulletin*, March 7. https://www.loc.gov/loc/lcib/94/9405/film.html.

Lomke, Evander. 2013. Psychiatry films from AMHF: "One Flew Over the Cuckoo's Nest" (1975). American Mental Health Foundation, January 9. http://americanmentalhealthfoundation.org/2013/01/psychology-films-from-amhf-one-flew-over-the-cuckoos-nest-1975-2/. Accessed 24 Feb 2018.

Martinez-Leal, R., L. Salvador-Carulla, C. Linehan, P. Walsh, G. Weber, G. Van Hove, T. Maatta, B. Azema, M. Haveman, S. Buono, A. Germanavicius, H. van Schrojenstein Lantman-de Valk, J. Tossebro, A. Carmen-Cara, D. Moravec Berger, J. Perry, and M. Kerr. 2011. The impact of living arrangements and deinstitutionalization in the health status of persons with intellectual disability in Europe. *Journal of Intellectual Disability Research* 55: 858–872. https://www.ncbi.nlm.nih.gov/pmc/articles/PMC3166640/.

Milos Forman: Milos Forman's Official Website, 2017. https://milosforman.com/en/movies/one-flew-over-the-cuckoos-nest. Accessed 24 Feb 2018.

Moffic, H.S. 2014. We are still flying over the cuckoo's nest. *Psychiatric Times*, July 1. http://www.psychiatrictimes.com/apa2014/we-are-still-flying-over-cuckoos-nest.

Native American Indian Association of Tennesee. On Use of the Title 'Chief': Native American Protocol 2017. http://www.naiatn.org/about/media-info/on-title/. Accessed 19 Jul 2017.

Navasky, Miri, and Karen O'Connor. 2005. *The New Asylums.* Frontline, PBS, released May 10. http://www.pbs.org/wgbh/frontline/film/showsasylums/.

Novella, Enric J. 2010. Mental health care and the politics of inclusion: A social systems account of psychiatric deinstitutionalization. *Theoretical Medicine and Bioethics* 31: 411–427.

Oregon Public Broadcasting System. 2005. Ken Kesey. http://www.pbs.org/opb/thesixties/topics/culture/newsmakers_3.html. Accessed 24 Feb 2018.

Oregon State Hospital Museum Project. 2012. https://oshmuseum.wordpress.com. Accessed 24 Feb 2018.

Rich, Nathaniel. 2012. Ken Kesey's wars: "One Flew Over the Cuckoo's Nest" at 50. *The Daily Beast,* July 26. http://www.thedailybeast.com/articles/2012/07/26/ken-kesey-s-wars-one-flew-over-the-cuckoo-nest-at-50?source=dictionary.

Roth, Loren, Alan Meisel, and Charles W. Lidz. 1977. Tests of Competency to consent to treatment. *American Journal of Psychiatry* 134: 279–284.

Shatkin, Jess P. 2013. The History of Mental Health Treatment. Presentation to the AACAP annual meeting, Orlando, FL, October 22–27.

Swaine, Jon. 2011. How 'One Flew Over the Cuckoo's Nest' changed psychiatry. *The Telegraph,* February 1. http://www.telegraph.co.uk/news/worldnews/northamerica/usa/8296954/How-One-Flew-Over-The-Cuckoos-Nest-changed-psychiatry.html.

Sutherland, Zena. 1990. *Orchard Book of Nursery Rhymes.* New York: Orchard Books.

TES Global, Inc. 2017. Social Movements of the Sixties: Native American. https://sixties-social-movements-3.wikispaces.com/Native+American. Accessed 24 Feb 2018.

The Zinn Education Project. 2017. Native American Activism 1960s-Present. https://zinnedproject.org/materials/native-american-activism-1960s-to-present/. Accessed 24 Feb 2018.

Thomas, Dexter. 2016. Meet the woman who inspired Marlon Brando's Oscar and inspired Jada Pinkett Smith's boycott. *Los Angeles Times,* February 5. http://www.latimes.com/entertainment/movies/moviesnow/la-et-mn-sacheen-littlefeather-oscars-20160204-htmlstory.html.

Torrey, E.F. 1997. *Out of the Shadows: Confronting America's Mental Illness Crisis.* New York: Wiley.

United States Senate, Ninety-Fifth Congress. 1977. Project MKUltra. Senate Intelligence Report. U.S. Government Printing Office. https://publicintelligence.net/ssci-mkultra-1977/. Accessed 24 Feb 2018.

University of Groningen. 2012. The Native American Movement. http://www.let.rug.nl/usa/outlines/history-1994/decades-of-change/the-native-american-movement.php. Accessed 24 Feb 2018.

Vallance, Tom. Dale Wasserman: Playwright who adapted 'One Flew Over the Cuckoo's Nest' for the stage. 2009. *Independent*, January 7. http://www.independent.co.uk/news/obituaries/dale-wasserman-playwright-who-adapted-one-flew-over-the-cuckoos-nest-for-the-stage-1229960.html.

Waldman, Amy. 1999. Iron Eyes Cody, 94, an actor and tearful anti-littering icon. *New York Times*, January 5. http://www.nytimes.com/1999/01/05/arts/iron-eyes-cody-94-an-actor-and-tearful-anti-littering-icon.html.

Wicker, Tom. 1975. *A Time to Die: The Attica Prison Revolt*. New York: Quradrangle/The New York Times Books Company.

Wiseman, Frederick. 1967. *Titicut Follies*. Distributed by Zipporah Films, Inc, released October 3.

Wolf, Tom. 1968. *The Electric Kool-Aid Acid Test*. New York: Farrar, Straus and Giroux.

Zipporah Films, Inc. 2018. Description of Titicut Follies. www.zipporah.com/films/22. Accessed 24 Feb 2018.

Dementia and Capacity: Still Alice (2014)

7

Still Alice is a gripping film about early onset dementia and the gradual loss of autonomy and decision-making capacity. It is the ideal film for any neurology or gerontology unit because it confronts everyday issues in these fields. The film challenges conceptions of personhood, and when we determine that someone is non-autonomous, as there is no bright line in this portrayal—just as there is no bright line in many cases. The film also raises questions about quality of life, as the character, Alice Howland, has a fit, healthy body; just not a sound mind, which makes us re-think traditional concepts of "end of life". This film is also about family systems (discussed more in Chap. 3 on *My Life*), and explores how loss of a family member's autonomy affects the rest of the family unit. The issue of genetic testing in family members is particularly explored, as Alice discovers her dementia is genetic, and probably affected her father (who she assumed died of alcoholic-related dementia). The film raises hard questions about the limits of advance directives, and the role of surrogate decision-makers and caregivers. At its heart, this film is about the nuances of decision-making capacity, which we see waxes and wans in a sea of "grey" as the character pops in and out of lucidity.

Origins of *Still Alice*

This film is based on a fictional book written by Lisa Genova, born in 1970, who became a novelist after completing her Ph.D. in neuroscience from Harvard University in 1998, and spending time as a researcher at the NIH. The author has been compared to Oliver Sacks as an informant of neurological tales. Genova was inspired to write the book based on her family's experiences with her 85 year-old

© Springer International Publishing AG, part of Springer Nature 2018
M. S. Rosenthal, *Clinical Ethics on Film*,
https://doi.org/10.1007/978-3-319-90374-3_7

grandmother's decline from Alzheimer's disease. After she wrote the novel, she was unable to find a publisher (told that it had very narrow and limited appeal to only Alzheimer's disease sufferers) and took advantage of the popularity of self-publishing. Still Alice was originally published by iUniverse in 2007, and then re-released in 2009 by Simon and Schuster, after a popular review (Beckham 2008) caught the eye of the large, traditional publisher. In her personal life at the time, Genova was going through a shift; she was amidst a second divorce from her second husband, and feeling somewhat "lost" and a failure in her personal life. She had two small children at the time. Genova experienced the universal story of female writers thriving after finding their "voice" through their writing. The popularity of her self-published book landed her a six-figure book deal with Simon & Schuster, and Still Alice was on the *New York Times* Best Seller list for 59 weeks (Genova.com), where she states:

> This is a very strange career I've stumbled into…I never wrote a word in my life, except scientific stuff …[My books are] accessible stories about people living with neurological conditions who are ignored, feared, or misunderstood. I see the stories as a vehicle for empathy and social change.

Genova had crossover interests in both neuroscience and fine arts; she had taken acting classes in addition to her writing. The characters in Still Alice demonstrate Genova's tensions between her academic career (expressed through Alice) and her interest in the Arts (Alice's daughter).

Still Alice becomes known as a much-loved book during an election year in which former President Barack Obama and Hillary Clinton were locked in a battle for the Democratic nomination in a series of over 20 debates that primarily focused on their different visions of healthcare reform. It was also the year in which the U.S. experienced its worst financial crisis since the Great Depression. Healthcare issues for a tsunami of aging baby boomers facing their own health problems as well as the burdens of caregiving in a context of losing their life savings, was very much on the minds of readers at this time. As discussed further (History of Medicine), research for prevention and treatment is the most critical discussion surrounding this disease, and in 2008, there was no treatment, and research funding was not optimal. By 2009, when President Obama is inaugurated, the book is republished by Simon and Schuster and becomes a *New York Times* bestseller. There is a brief period in which the book is turned into a play; Christine Mary Dunford developed it for the Lookingglass Theatre Company in Chicago, and had a run April 10–May 19, 2013 (see: https://lookingglasstheatre.org/event/still-alice/).

The book unabashedly is a "call to arms" for donations into Alzheimer's disease research, evidenced by the author's website (see: http://lisagenova.com), and Genova was considered, by 2015, to be one the most influential individuals with respect to Alzheimer's disease research. More importantly, the book resonated strongly with readers regardless of whether they were personally touched by Alzheimer's disease; the book has a "can't put it down" quality to it. The book also won acclaim by neurologists with respect to its accuracy. One book reviewer notes at the time (Beckham 2008):

The monster in "Still Alice" isn't an invention. It's real. It's Alzheimer's disease and some 5 million people in the US are living with it, half a million younger than 65, just like Alice Howland, the fictional 50-year-old professor, whose world is all words and lectures and intellect. Alice lives with Alzheimer's disease and ignores it and battles with it and tries to tame it and outrun and outwit it….."Still Alice" is written not from the outside looking in, not from the point of view of a caretaker or a husband or a friend, but from the inside looking out. This is Alice Howland's story, for as long as she can tell it. …Every 72 seconds another American develops Alzheimer's disease. The literary agents who said Genova's book wouldn't sell are wrong.

The book is credited with beginning broad public discourse about Alzheimer's disease. There are some thematic similarities between *Still Alice* and *Wit*, which I cover more under "Women in academia" further on (under Social location). Both are experts in areas that their diseases challenge. In *Wit* (see Chap. 1), the main character, a 50 year-old professor, is an expert on John Donne's poetry about death (the Holy Sonnets), but has trouble dealing with her own death. In *Still Alice*, a 50 year-old professor is an expert in linguistics, but is losing her language skills. In *Wit*, the poetry of John Donne is a central theme that helps to feed the story; in the film, *Still Alice*, "Angels in America", a play by Tony Kushner, first performed in 1991, and which won a Pulitzer Prize for Drama in 1993 (Kushner 2013), is mentioned numerous times, as Alice reads lines from it with her daughter (who is preparing to act in the play). "Angels in America" is a play about the early days of AIDS, and how it affected patients and families; it was made into an HBO miniseries in 2003. Alzheimer's disease is similar to early AIDS in that it is a progressive disease that has no cure, and exacts the same type of social death that early AIDS did prior to AZT and the accompanying drug protocols. Depending on the curriculum *Still Alice* is part of, one might consider pairing this content with a study of AIDS, "Angels in America", and perhaps even *And the Band Plays On* (see Healthcare Ethics on Film) as a grouping.

The Film

The film rights were sought for the book, Still Alice, as early as 2008 (at first by an Alzheimer's disease patient advocate with a film background) but Genova waited until 2011 to sell them to a British filmmaking couple, James Brown and Lex Lutzus. Genova recalls: "It was a leap of faith on my part, but they just really understood the point of the story that had to be told from Alice's subjective point of view."

The Executive Producer is Maria Shriver, the daughter of Eunice Kennedy Shriver, and niece of John F. Kennedy. Shriver's father had Alzheimer's disease, but her mother, Eunice, had been the primary caretaker of Rosemary Kennedy, who had undergone a lobotomy at the request of Joseph Kennedy (father of Eunice and Rosemary). Eunice's life's work had been dedicated to disability rights, and influenced Maria similarly. Regarding *Still Alice*, Shriver states (Elwood et al. 2014):

I met [Genova] when she brought her book to The Women's Conference several years ago...My dad died of Alzheimer's disease and it's affecting people every 90 seconds in this country...My hope, having done a lot in this field through the news business, through books, through HBO documentaries, is that this film being brought to life by these incredible actors does something we haven't been able to do, which is to put it into the mainstream, to make young people become interested in this disease, because it will affect their parents and it's affecting women disproportionally at a very young age....

She also states (Farhi 2015):

A lot of people are comparing this to [the film,] 'Philadelphia';,,What 'Philadelphia' did for AIDS, 'Still Alice' can potentially do for Alzheimer's disease.

The film was written and directed by Richard Glatzer and Wash Westmorland, filmmakers who were married. Glatzer and Westmoreland were approached by their friends, Lutzus and Brown, to adapt the Genova novel. At this point, Glatzer had just been diagnosed with amyotrophic lateral sclerosis (ALS). The script was finished in 2012, and some of it was based on Glatzer's insights about his own neurological disease. When Glatzer met with Genova over dinner in September 2013 to discuss the film project, Genova was resolved to eventually write a novel about an ALS patient; she has so far tackled brain injury, autism and Huntington's disease in other novels. Ultimately, the film was independently produced on a small budget by Neon Park Productions and Killer Films, owned by Pam Koffler, who is credited as a third producer of the film. It was shot in 27 days in New York City; Alice and her husbands' workplaces were changed from Harvard University to Columbia University. In the film, John, leaves her in New York City with her daughter to live with her, while he accepts a Department Chair job at the Mayo Clinic. (In the book, John leaves Harvard for Memorial Sloan Kettering, but details about his leaving are unclear, since we are only privy to Alice's perspective—and by this time, she does not know who he is, or where she is.)

Julianne Moore was Glatzer and Westmoreland's first choice for the lead role. To prepare for it, Moore did several months of research, including meeting with several Alzheimer's disease patients, their caregivers and support groups, healthcare providers and patient-led organizations. Moore also underwent the same cognitive testing and dementia screening tests with a neurologist that Alice undergoes in the film. The film had several expert advisors, including the Alzheimer's Association Chief Science Officer, Maria Carrillo, Ph.D.; and the Alzheimer's Association National Early-Stage Advisor Sandy Oltz, a patient who was diagnosed with younger-onset Alzheimer's disease at age 46. Says Moore (Elwood et al. 2014):

What I think is so compelling about this movie for me is that it's really about who we are, essentially. There is a reason it's called 'Still Alice.' In the face of anything affecting your life in this way, who are you essentially? In the face of a terrible disease, who are you to your loved ones, your children, your husband, your job? Who are you to the world? It's not that someone disappears. I think it's a misnomer that someone with Alzheimer's disease goes away. The person is there, they are just there in different capacities. How do you express that? I think it's interesting in terms of our relationship with Richard and Wash that they were in a somewhat similar situation. The interesting thing about communicating with

Richard is that what he is going through kind of disappeared after awhile because you were still communicating with the person who is there. Oddly it wasn't an issue. So, in a way, it made it very clear to me what was the defining issue of the film.

Synopsis

Similar to *Wit* (see Chap. 1) this film is about another 50 year-old female professor-patient in a very different circumstance: Alice Howland is a linguistics professor at Columbia University who has just turned 50, and is beginning to notice problems with her memory. She seeks out a consultation with a neurologist and learns that she has a rare form of early onset Alzheimer's disease that is genetically inherited and autosomal dominant, in which the child has 50/50 chance of inheritance from the parent. The film is about her descent into dementia, including a botched attempt at suicide, and examines her disease from her perspective and her family's perspective, as her children need to make a decision about whether to get tested. Two of her children get tested; one of them declines. Alice's husband (played by Alec Baldwin) is a well-established physician and basic researcher, and they have three grown children in their 20s. Pre-dementia, Alice struggles the most with her youngest daughter (Kristin Stewart), who chooses acting over college. Ultimately, Alice's husband is unable to cope, and makes a decision to take a new job at the Mayo Clinic in Minnesota, and arranges for Alice to remain in their New York City home; the actor-daughter elects to return to live at home as the primary caregiver and family member in charge of Alice. We learn an interesting fact about dementia in the film: the smarter and more educated you are, the faster the seeming decline. This is because the smarter you are, you can remain functional longer, and you have learned to compensate or conceal your symptoms. Alice, as it turns out, has been experiencing symptoms long before she seeks help, and is diagnosed when her dementia is far more severe.

2007–14: Seven Years of Recent Memory and the Social Location of *Still Alice*

Unlike many films discussed in this book, *Still Alice* is essentially a book and film that is being developed within a social milieu most viewers will remember. The book, originally self-published in 2007, comes out in the end of the Bush years, but at a time when a major film by Michael Moore, *Sicko* is released (see Healthcare Ethics on Film). *Sicko* provides a grim diagnosis of the American healthcare system as baby boomers (1946–64) begin to age and begin to deal with caregiving of their own aging parents. Lisa Genova, as noted earlier, was inspired to write the book based on her family's experience of caring for her 85 year-old grandmother with Alzheimer's disease. Genova is not a baby boomer; born in 1970, she is considered to be part of

Generation X (1961–81), which some tail-end boomers identify with more. Genova's parents, however, are classic baby boomers who were challenged with caregiving. Moore's film, *Sicko,* is what triggers a major debate over healthcare reform during the 2008 election. From January through June of that year, Barack Obama and Hillary Clinton were locked in a tight race for the Democratic nomination and debated over 20 times; they mostly used these debates to carve out different solutions to healthcare reform overhauls: Obama championed the "Romneycare" model, while Clinton championed a single-payer system. By 2009, when Obama (born in 1961) wins, he represents two generations: the tale end of the boomers, and, according to some statisticians, the beginning of the Generation X cohort. The Obama Era (2009–17) is considered to be the most optimistic political timeframe in the United States that coincides with the end of "post-9/11". However, the first term of Obama is in "rescue mode" as he inherits a devastating financial crisis that began in the Fall of 2008. Genova's tale of memory loss is overlaid onto a larger social context of financial loss for the country—a timeframe in which thousands of jobs are lost; billions in savings are lost; and Americans begin to feel as though their identities are becoming "lost" as well, as they lose their homes, jobs, savings, and connections to who they were prior to the financial crisis. This period in the Obama era is focused on recovery, and fierce debates in Congress over austerity leads to enormous cuts in research funding at the National Institutes of Health. But as baby boomers and their parents' generation ages, Alzheimer's disease accelerates, and Genova serves as a conduit for awareness and lobbying for funding into the disease.

In 2008, a major celebrity, Charleton Heston, who was also President of the NRA (1993–2003), passed away from Alzheimer's disease; he was diagnosed in 2002, and announced it publically to his fans in a 2002 letter before retiring from public life a year later—similar to what Ronald Reagan had done in 1994, when he was diagnosed (Madigan 2002).

In 2011, news stories begin to emerge surrounding former President Ronald Reagan, who died from Alzheimer's disease complications in 2005. We learn that Reagan struggled with serious dementia most of his second term, which had been hidden from the public; the source of much of this information is from his own son, Ron Reagan, Jr., whose book about his father reveals this. An HBO documentary is released in 2011, *Reagan*, which provides a much more in-depth analysis of Reagan's struggle with dementia. A poignant ending tells the story of Reagan picking up a glass "snow globe" in his home with the White House inside it. He says to Nancy Reagan: "I know this house, and that it has something to do with me, but I can't remember what it is." The story leaves every viewer of the documentary —despite political leanings—in tears.

Another major event takes place that same year: the end of the Iraq War (2003–11). After a long conflict in Iraq that seemed to have no exit strategy, Obama makes the decision to follow a Bush-era deal made to hand sovereignty back to the Iraqi government and pull out all remaining American troops, many of whom are suffering from the signature wound of the Iraq war: traumatic brain injury (TBI), of which memory loss and some shared features of dementia begin to impact veterans and military families. Genova's character, experiencing neurological symptoms as a younger patient, resonates with a new generation of patients experiencing TBI (see under History of Medicine). *Still Alice* also resonates with an unlikely population: football players. By 2013, news emerges of the NFL's cover-up of football-related dementia, and severe brain injury from concussions. We learn in 2013, that this is a decade-long cover-up in which football players begin to complain about the impact of their post-football neurological states. As early as 1993, Dallas Cowboys quarterback, Troy Aikman, who suffers a head injury, tells a journalist that he has no recollection of the game and had lost his memory. Another player, Mike Webster files a disability application with the NFL Retirement Board in the late 1990s, claiming his NFL football career "caused him to have dementia". Coinciding with Genova's novel, studies begin to be published linking concussions to dementia through a condition known as Chronic Traumatic Encephalopathy (CTE) (Fainaru-Wada and Fainaru 2013; Kirk et al. 2013). I discuss these issues more and the film, *Concussion* (2015) in Healthcare Ethics on Film.

By 2014, when *Still Alice* is released as a film, interest in dementia and Alzheimer's disease research spikes as the details of Robin Williams' suicide emerge; Robins had hung himself after being diagnosed with a form of Parkinson's disease and learning that he will succumb to dementia. Although Williams' struggled with drug addiction and bipolar depression, it was upon learning he had dementia (as part of his Parkinson's disease) that led him into making a decision to end his life while he still had capacity (Smith 2015). The film garners rave reviews as well as more attention to early onset Alzheimer's disease (LeMire 2014; Seymour 2015; English 2016; Learner 2016).

The Role of Technology

In the novel, which takes place between 2003 and 2005 (though published in 2007), the role of Alice's Blackberry is omnipresent, as it helps her with her symptoms. In 2007, the iPhone is introduced, which becomes the dominant smartphone in the market. The film features Alice's iPhone instead of a Blackberry—again, as a major tool in helping her remember and organize and deal with her symptoms. It is worth discussing how such tools have become an important part of an Alzheimer's disease patient's world, and how technology has helped to improve quality of life.

Women in Academia

Lisa Genova made her protagonist a world renowned linguistics professor at Harvard (novel), which was changed to Columbia University for the film. Since Genova's own field was neuroscience, this is likely the reason she has built the story around an academic, perhaps mirroring some of Genova's observations about women in academia. (It is possible, too, that Genova may have known about *Wit*, and decided to use an academic character as well.) When Genova is working on her novel, there are many issues surrounding gender inequities in academia—particularly in traditionally male-dominated professions, such as the sciences, or "STEM" (Science, Technology, Engineering, Medicine). In 2009, a Census Bureau's

American Community Survey found that although women comprised 48% of the workforce, only 24% were in STEM careers (Beede et al. 2011). In academia, women were found to be consistently paid lower, promoted less often, and did not occupy leadership roles such as Chairs or Deans (Misra et al. 2011). This began to shift, but Alice Howland is representative of a woman who does become a full professor with tenure at a major academic institution, only to have to give it all up. Symbolically, women struggled to have their intellects valued as much as men; Alice's dementia in the context of a "brainy woman" may be Genova's nod to the unintended curriculum for women. This may resonate quite deeply with female faculty colleagues, as the academic work/life balance of this couple is familiar. Alice Howland, we learn, has had to juggle her career and children, like many women, but her husband, John, rises faster as a result of the fact that he does not need to take time off for raising children, or extend his tenure "clock". Nonetheless, we meet Alice at 50 at the peak of her career, only to watch everything collapse; Alice has succeeded in "having it all"—as some women do—and in this context, is she being "punished" on some level?

She discloses her disease to her Chair after learning that her course evaluations are poor, and begins to face the loss of her academic life, which is so much a part of her personhood. The descriptions of academic life in both the novel and film are very realistic. It should be noted, here, too, that we are still viewing a diagnosis of dementia in the context of "white privilege"—a couple that has insurance and connections to diagnosis and treatment, with access to genetic testing to boot. From a cultural competency standpoint, it would be important to discuss how dementia might be different for families that have absolutely no access to a diagnosis, and certainly no access to treatment or knowledge of clinical trials (see under History of Medicine below).

History of Medicine

Alzheimer's disease (AD) was first described in 1906, named after Alois Alzheimer's disease, who had a patient ("Auguste D.") with severe memory loss, and the typical psychological changes; but Dr. Alzheimer did an autopsy on his patient and found shrinkage around the brain's nerve cells, which was uniquely observed at that point. When the electron microscope was invented in 1931, the brain cells of these patients could be studied further. Cognitive testing was developed in 1968, which was refined over the years, but diagnosis is essentially a clinical diagnosis aided by cognitive tests. Between 1974 (when the National Institute on Aging or NIA is established) and 1993, gene mutations were identified for some forms of Alzheimer's disease (see further) and allocation of resources for further research, led to the first not very effective drug in 1993 (Cognex). In 2003, further research into genetic markers for AD speeds ahead, because it yielded the most promising actionable steps in prevention with the availability of genetic screening. In retrospect, this disease has few historical milestones in terms of treatment for current patients; drugs to slow disease progression generally are not very effective in the long term. When teaching, there are three history of medicine contexts to discuss with this film. First, a review of early onset Alzheimer's disease during this time frame is important.

The second context to discuss is the genomics context. *Still Alice* takes place in a post-human genome era, a timeframe where whole genome/exome sequencing is starting to become more widely accessible to patients, but also a time where the FDA clamps down on direct-to-consumer genome companies such as 23andMe (Brandom 2013; Bailey 2013). The book and the film take place within the context of the 2008 passage of the *Genetic Information Nondiscrimination Act* (GINA), which took place during the Bush Administration (NHGRI 2008; EEOC 2008; U.S. Congress 2008). Within this context, genetic testing is presented as an important step in early diagnosis, starting early mitigating therapies, and even preventative treatment in the next generation using preimplantation genetic diagnosis (PGD) (see further). But the film confronts the ethical downside of genetic screening for at-risk family members as well (see under Clinical Ethics Issues). It's important to point out, too, that the film is released just a year before the major scientific breakthrough of gene editing using CRISPR-cas9, but the same year Moore wins her Oscar for best Actress in the film (February 26, 2015).

A final context to discuss are the drugs used in slowing the progression of Alzheimer's disease, as well as health disparities in Alzheimer's disease diagnosis and treatment.

Early Onset Alzheimer's Disease

Still Alice is about early onset Alzheimer's disease (aka "younger onset"), which affects people younger than 65, representing only about five percent of Alzheimer's disease patients, which in the U.S., accounts for about 200,000 patients. In these cases, patients begin to show symptoms in their 40s and 50s (Alzheimer's Association 2017a). Alice is 50 when we meet her, but as we learn in the film (and novel), she had likely been symptomatic for much longer, but it was ignored. It's also important to note that in many older dementia patients without any access to treatment, some may have suffered from undiagnosed early onset. However, even in patients with healthcare access, most primary care physicians may not recognize signs of early onset Alzheimer's disease because symptoms can be attributable to multiple causes such as sleep deprivation, stress, and in women, menopause, which is what Alice initially believes herself. Thus, a typical experience with early onset Alzheimer's disease is delayed diagnosis, which can interfere with patient decisions when time is of the essence. As the film also relays, a diagnosis may not be conclusive without an autopsy, as frequently brain imaging scans may not detect early onset; in Alice's case, her brain imaging tests are negative. Thus, the only way to diagnose it in some cases, is based on clinical presentation of memory and cognition problems. Alice ignores her symptoms until she has an episode of complete disorientation after running in her neighborhood, and not knowing where she is. All of her other symptoms—forgetting words, absent-mindedness to a more severe degree, and so forth, are rationalized as "stress, aging, menopause" and so forth. The novel goes into a lot of details about her cognition testing, and the film less so, but the film does show an accurate representation of a neurology visit and testing. (Not as detailed as in a real simulation, but a good representation.) Accurate diagnoses of early onset Alzheimer's disease still rely on a neurological exam and cognitive tests; genetic testing can confirm a diagnosis only in cases where there is a genetic component (see further). Not all cases of early onset Alzheimer's disease are genetic; they may be idiopathic. Alice's case is a diagnosis of early onset Alzheimer's disease in a later stage due to the rapid decline she experiences from her date of diagnosis to a dramatic decline in functioning (Alzheimer's Association 2017a), which seems to occur rapidly within about a year. As her neurologist accurately explains, Alice's intelligence succeeded in masking symptoms until a much later stage, when Alice finally sought out help for what clearly could not be attributed to normal stress (finding herself lost in her own neighborhood for a good ten-minute stretch of time). Memory loss that disrupts daily life may be elusive, especially with smart phones and instant access to the internet, GPS devices, and so forth.

In terms of diagnosis, not much has really changed since the film was released in 2014 as of this writing, but if you view the film when limitations are indeed historical, it's important to discuss what has changed. When the film is released, the testing for Alzheimer's disease and dementia involve a primary care visit that exhausts other causes, such as mental status and mood disorder evaluations, and a thorough physical. The next step would be a thorough neurology exam, similar to what is presented in the film, as well as cognitive tests (neuropsychology tests) blood tests and brain imaging tests to rule out things such as brain tumors, and in very advanced cases, brain atrophy can be seen on an MRI. There are no protein markers in Alzheimer's disease. Genetic testing would be recommended in cases where dementia is diagnosed. Genetic testing can reveal a variety of genomic circumstances that could be useful to at-risk family members. Looking forward to a post-gene editing context, there could be clinical application of gene editing in the future for Alzheimer's disease patients and at-risk family members, but that is years away from application, and may have many ethical implications (see further).

There are other causes of early dementia that may include head injuries (such as those seen with traumatic brain injury or CTE, discussed earlier); dementia may also be caused from decreased oxygen to the brain due to a cardiovascular problem ("vascular dementia"). In Alice's case, she has familial early onset Alzheimer's disease caused by an autosomal dominant gene she presumes has been passed down from her father, whose dementia was unfortunately masked by his chronic alcoholism. Alice reports her father was completely "out of it" by the end of his life, but presumed it to be connected to his drinking. This presentation is not unusual in many families—particularly in an era where addiction is so prevalent.

The film also does a good job of presenting the limitations of treatment; Alice is on Namenda and Aricept, and other nutriceuticals, but nothing seems to help, and this is consistent with the literature on the poor efficacy of AD drugs; the current controversy in drug therapy for AD lies in why some of the drugs are promoted to patients and families in the first place, and whether they actually meet the criteria for "beneficence" (see next section). There are questions surrounding inflated results, insufficient rigor in clinical trials, and therapeutic misconceptions about false hope (Casey et al. 2010). Essentially, the film presents the "future" as the best treatment: genetic screening for at-risk family members; prenatal genetic diagnosis (PGD) to "stop the gene" from carrying forward into the familial lineage; and entering a clinical trial, which would mostly be of use to a future patient. I discuss clinical trials in more detail in Chap. 10, on *Awakenings*.

Pharmacotherapy for Alzheimer's Disease

Since the film was released and as of this writing (2017), there are five ineffective therapies for Alzheimer's disease that are FDA-approved (Casey 2010). They are so ineffective, in fact, that the United Kingdom's National Health Service stopped recommending most of them because the cost is not worth it. Four of these medications are classified as cholinesterase inhibitors (CIs), which are approved for mild to moderate dementia; they comprise tacrine (Cognex), donepezil (Aricept), rivastigmine (Exelon), and galantamine (Razadyne). Alice is on Namenda and Aricept, which is also FDA-approved for severe or late-stage dementia. No one uses tacrine because of its liver-toxicity side effects; a fifth drug approved, memantine (Namenda, Forest), blocks NMDA receptors, but is also not that effective (Casey et al. 2010).

A scathing analysis of drug therapy for AD states (Casey et al. 2010):

> The neurotransmitter effects of the CIs and the NMDA antagonists do not change the underlying brain degeneration characteristic of AD. The drugs do not seem to affect life span or outcomes of the disease. Therefore, they are best viewed as palliative rather than curative or disease-modifying treatments...A discussion of the differences between the concepts of efficacy and effectiveness may shed light on the controversies surrounding the use of drugs for AD. Efficacy is essentially a statistical concept; it is measured in placebo-controlled trials by demonstrating a statistically significant superiority of an active treatment over placebo and it uses a predetermined set of validated measures. A finding of statistical significance means that the result is likely to represent a true observation, as opposed to a coincidence. All current AD drugs have met this standard on multiple trials and therefore may be described as being efficacious...A finding of statistical significance does not necessarily imply that the magnitude of the effect, however real, is sufficient to justify the expense or risks involved.

> In treating a devastating and incurable disease such as AD, should we offer expensive but limited therapies?...Alzheimer's disease is an incurable neurodegenerative disorder that robs its victims of memory, self-care, and quality of life, resulting in major physical, emotional, and economic burdens for caregivers. These burdens must be considered on both societal and personal levels, because AD exacts an enormous financial drain on the medical system. Much of this cost consists of drugs that have some efficacy but limited effectiveness for most patients....Some critics, especially those concerned with conserving scarce resources, have claimed that these medications are not cost effective enough to justify the expense.

And yet, compare this to what patients read on the website, Alzheimer.org, which makes drug therapy sound promising:

> While there is no cure, prevention or treatment to slow the progression of Alzheimer's disease, there are five prescription medications approved by the U.S. Food and Drug Administration (FDA) to treat its symptoms...[C]holinesterase inhibitors... treat symptoms

related to memory, thinking, language, judgment and other thought processes. [M]emantine, regulates the activity of a different chemical messenger in the brain that is also important for learning and memory. Both types of drugs help manage symptoms, but work in different ways.

Ultimately, the drugs don't work well on Alice, and don't work well on anyone else, either. Alice, like many other patients, is also on Lipitor (to avoid vascular dementia), vitamins C, E and aspirin. Unfortunately, she likely benefits most from these than her dementia drugs.

Genomics Issues

In Alice's case, she has the equivalent genetic time bomb as Huntington's disease; she tests positive for one of the autosomal dominant gene mutations that is 100% penetrant. Unlike familial breast cancer, for example, familial Alzheimer's disease is 100% guaranteed to develop in those who test positive for the mutation. (Genova 2009). When teaching, it may be worthwhile to walk students through basic genomic concepts (see Table 7.1). It's important to get across when teaching that Alice tests positive for the Presenilin-1 (PS1) mutation, identified in 1992, which results in an autosomal dominant disorder that is 100% penetrant (Genova 2009). This means that genetic testing is the only pathway to successful prevention for Alice's at-risk children and their descendants. For Alice's children, knowing if they're positive early will help them recognize symptoms early, and potentially be able to take advantage of future effective therapies, when available. But they can also choose PGD, which entails conception through IVF methods so that DNA testing of the embryos can be done, and only the mutation-free embryos are implanted, thus ending the inherited mutation in the familial line. The mutation for early onset Alzheimer's disease represents a genetic disease where genetic testing offers definite benefit. In Alzheimer's disease, other mutations with 100% penetrance, called in the patient literature "deterministic" (Alzheimer's Association 2017a), includes amyloid precursor protein (APP) and presenilin-2 (PS-2). There are other mutations that increase the risk of Alzheimer's disease with lower penetrance (called "risk genes" in the patient literature), such as

Table 7.1 Basic terms in genomics

- Autosomal recessive: Two abnormal genes (one from each parent) needed for trait to be expressed
- Autosomal dominant: One abnormal gene (from 1 parent) needed for trait to be expressed
- Genotype: What the gene looks like
- Phenotype: What the *person* with the gene looks like, or how the gene is "expressed"
- Penetrance: If genotype is expected to produce a phenotype, what.% of the time will that occur?
- Autosomal recessive: 25% chance of offspring having gene if both parents have the gene
- Autosomal dominant: 50% chance of offspring having gene if one parent has the gene

apolipoprotein E-e4, or APOE-e4. In these cases, Alzheimer's disease runs in the family, but the likelihood of it occurring is less (Alzheimer's Association 2017a).

It's also important to review how genetic screening for a specific gene may become obsolete in the context of whole genome/exome screening, which is quickly becoming more accessible. In 2013, when 23andMe launched, it made genome screening and results available to consumers for $99.00. Due to concerns over how results could be interpreted, and potential harms from misinterpretation, the FDA cracked down on the company, restricting it to ancestry only and not health-related results. This recently reversed (Mullin 2017; Pollack 2015). Since the film was released, it is starting to make less sense to be tested for a single mutation when the entire genome can be screened, providing more information about health risks and early prevention.

Although I do not go into sufficient detail here, it would be important to discuss what occurs in early onset patients who have no access to primary care, screening or diagnosis in populations where neurological exams are completely unavailable. It is unknown, for example, how many individuals remain undiagnosed.

Preimplantation Genetic Diagnosis (PGD)

When *Still Alice* premiered as a film, PGD was a technique already in use for several adult onset diseases, and had been available for years (Handyside et al. 1990), but had been misused for sex selection primarily. In 2013, the American Society for Reproductive Medicine Ethics Committee endorsed the use of PGD for adult onset disorders in its 2013 Opinion under certain conditions, which would certainly include any degenerative disease for which there is no cure (ASRM 2013). In this sense, PGD can effectively be used to breed out known genetic diseases. It may be worth discussing with students whether this is also a form of justified "eugenics" as opposed to unjustified uses of PGD, for trait selection or sex selection, for example. Even in the context of ethically justified uses, we still need to remember that Alice was a world expert in her science and made major contributions prior to her disease onset; would we want to remove future "Alice Howlands"—who, while succumbing to dementia in mid-life, made major contributions in early life, including inspiring many graduate students? Would we be breeding out future "Einsteins" inadvertently? (Rosenthal 2015).

With respect to PGD, we also have to look at health disparities, and whether such technologies may lead us to a "Gattacan" world, in which the poor have diseases, including genetic diseases, such as this. It's certainly easy to see the benefits of preventing a harmed life; some bioethicists have argued that in select cases, it may even be "unethical" to have a baby if you're knowingly passing on a terrible disease. The President's Commission on Bioethics provided an opinion on PGD in 2004 (President's Council on Bioethics 2004), which stands the test of time:

> Biology is not destiny...The ability to affect the genetic make-up of the next generation
> may also exacerbate the tendency to assign too much importance to genetic make-up, and

so may promote an excessively reductionist view of human life. These new practices may lend undue credence to the notion that human characteristics and conditions are simply or predominantly genetically determined—a too-narrow understanding of human freedom, agency, and experience, and a simplistic understanding of human biology.

Gene Editing

A few months after *Still Alice* premiered, fears of human genome editing were simmering by Spring 2015, after multiple studies were published by various research groups around the world surrounding "gene editing" using the technology CRISPR-Cas9 (Yin 2015). In response, there was a call for a moratorium on gene editing of human germ lines because there were concerns that the science was ahead of the ethical implications (Lanphier et al. 2015; Baltimore et al. 2015). But it was too late; a Chinese team of researchers published in April 2015 that they had performed such an experiment on human embryos (Liang et al. 2015). This created raging debates regarding the ethical implications of the experiment, which boiled down to fears that we would be using the technology in a "Frankenstein" type of fashion, with unintended consequences (Cyronski and Reardon 2015; Kolata 2015). The first summit on human gene editing was held December 1–3, 2015 (Maron 2015), which led to moral consensus on three key issues:

- Refrain from research and applications that use modified human embryos to establish a pregnancy.
- Caution in development of clinical applications until we know more about safety and efficacy, and the risks of inaccurate editing.
- Moratorium on 'germline' editing—the deletion of a gene prenatally in an effort to erase an inherited disease from an embryo and prevent it from being passed on to future generations.

It's important to raise with students that the main ethical concerns with gene editing is whether we are eliminating "defects" from the genome that are, in fact, attributes, or introducing "attributes" that wind up changing the human condition in unfathomable ways. Notwithstanding, most people would want to prevent early onset Alzheimer's disease, and it would be a defensible application of this technology.

Clinical Ethics Issues

The overarching clinical ethics issues in this film revolve around the loss of autonomy and diminishing or evolving decision-making capacity in a dementia context. Other issues raised in this film surround disclosure of the illness; advance directives and their limitations in a dementia context; assisted suicide; and ethical issues with genetic screening decisions. Finally, Alice's family is profoundly affected by her disease, which raise important themes about "family systems" (see Chap. 3 on *My Life*). You may also want to review some less prominent clinical ethics issues in the film, such as the truth-telling scenes between Alice and her neurologist as well as the discussion of clinical trials raised in the film. For a more in-depth discussion surrounding clinical trials, I refer readers to Chap. 10, on *Awakenings*.

Autonomy and Decision-Making Capacity in Alzheimer's Disease

Whenever I teach this film, the first question I ask is: "At what point in the film does Alice lose decision-making capacity?" Or, "At one point does she lose her autonomy?" No one can quite answer these questions because both autonomy and decision-making capacity are fluid and dynamic in the case of dementia. I have previously discussed the distinctions between competency and capacity (see Chap. 6). Here, we will focus solely on *decision-making capacity*, and the U-ARE criteria, which stands for Understanding, Appreciation, Rationality and Expression of a choice (see Table 5.1).

I argue that we need to look at the separate preferences expressed by the pre-dementia, and post-dementia self in terms of guiding medical decisions, and *Still Alice* nicely demonstrates this. To that end, it is unclear what the status of Advance Directives may be in these circumstances as well, other than to select a surrogate. The most poignant example in the film about the "grey areas" of decision-making capacity and the role of Advance Directives is the "Butterfly File" scene and the accompanying "botched suicide" attempt later in the film. When Alice is still fully capacitated in the first half of the film, she constructs a "self-test" that would provide a bright line, for her, that she has lost capacity and autonomy. She constructs a daily test of questions to answer (such as her address; her children's names and birthdays), a self-test that tells her what to do when she can no longer answer them. Alice rationalizes that when she can no longer answer these questions, life wouldn't be worth living, and that suicide/death with dignity would be the optimal solution at that juncture. At this stage of her illness, she acknowledges to her family that she is not so much "suffering" but "struggling". So her still-capacitated self instructs her future *incapacitated* self to go to a "Butterfly" file on her computer if she can't answer the questions anymore. The "Butterfly" file is a folder named "Butterfly" (this is so-named for a favorite piece of jewelry shaped like a butterfly from her mother), and when clicked, a video runs, which she had made that talks to her future self. It instructs her to go to her bedroom night table and swallow all the pills in a bottle labeled "ALICE" that she has tucked away; it is a bottle of sleeping pills that Alice had secured when she still had capacity. The plan sounds perfectly logical. Of course, it relies on the fact that Alice would remember to take her daily test and appreciate she has the wrong answers (which she doesn't). None of this goes according to plan exactly; Alice finds the

"Butterfly" file when searching for something else on her computer while flustered, and really does try to follow the directions on the video, but gets so distracted, she cannot not follow through; she finds the pills but drops and scatters them everywhere. In about five minutes, it's as though she never watched the video, and the carefully planned Advance Directive Alice left for herself is gone. The viewer realizes that "life is what happens in between making plans"; indeed, all the planning of Alice's capacitated self falls away because she doesn't account for the fact that she won't be the same person when it's time to follow the directions.

Alice's husband, who is unaware of her Advance Directive, brings up euthanasia to Alice, too, in the "ice cream" scene. In the film, when Alice is considerably compromised cognitively, John takes her to their favorite ice cream shop and they sit together eating their ice cream in full view of Columbia University, where Alice used to work. John points out the familiar landscape and asks her if any of it looks familiar, or if she remembers she used to work there. Alice does not remember the location but knows she "used to be very smart". John pauses and says: "Do you still want to be here?" We all know what he means, and Alice's previous self would have understood and appreciated that statement, too. She replies: "I'm not done yet; do we have to go?" making clear that she's not finished with her ice cream. Right there is the bright line that separates Alice's past and present self. The previous self had memory that defined them, and informed depths of existential context. Many Alzheimer's disease patients and anyone suffering from dementia still live fully— but only in the present, and sometimes only in the moment. Alice, post-dementia still has quality of life; she enjoys her food, can use her healthy body, appreciates the sights and sounds of her surroundings, and, can still appreciate a moment. She wants to stay to finish her ice cream, and that is still a preference that indicates she is still busy living and not thinking about dying. Her previous self passed judgment on what late-stage dementia would feel like. Now that she is in late stage dementia, she is just "being". John drops the discussion, realizing that while it's true that Alice clearly didn't understand and appreciate what he actually meant (*does she want him to help her to die?*), she really is enjoying her ice cream, and expressing her preference to *stay*. This scene harkens the adage, "the future's a mystery, the past is history, and the present is a gift." Alice still controls some of her destiny. Therein is also the guidance I typically provide surrounding how to deal with decision-making capacity with dementia: it is not all or nothing; it is task-specific. Patients can still make many decisions. Since Advance Directives typically do not "kick in" until a patient is unable to express wishes anymore in the context of a terminal illness, what do we do in these circumstances? If the pre-dementia autonomous self wants us to stop all feeding of the post-dementia self, should we do it? Similarly, if there is an Advance Directive in a patient with no dementia, who changes her mind about being "DNI" (Do Not Intubate) in an emergency—realizing she really does want a tracheostomy after all—we honor the most recently stated preference, which supersedes the previous directive. Similar questions have been raised in dementia surrounding a very common ethics consult in nursing homes: two dementia patients engage in a sexual relationship having forgotten they are married to other people.

There is consensus that healthcare providers should *not* interfere and police these situations, because although they have diminished capacity, they are still "consenting adults". Generally, if there is a firm Advance Directive written by the pre-dementia self, we would meet the patient "where she is" and review basic preferences at this juncture. Typically, many patients with dementia have quality of life, but a very different quality of life than what their pre-dementia self would find acceptable. This is not different than a quadriplegic or ALS patient adjusting and deciding that there really is something left to live for. A hard question to address is whether loss of memory and sense of the past self represents loss of personhood? I would argue it does not, given that we grant personhood status at birth. Since there are also moments of lucidity, and clear preferences stated, the clinical ethics view would be to continuously strive to guide care around last stated preferences, and to look for consistency. Alice's preference to "stay" thus prompts us to ensure that she would have a safe environment and adequate caregiving. Had she, alternatively, stated that she doesn't want to live anymore or "can't do this anymore" or had begun to stop eating, invoking VSED (voluntary stopping of eating and drinking), then that would guide us into respecting these wishes, too. Ultimately, decision-making capacity to consent may change to assent and preference-stating with details of care decided by others. Thus, the most important decision for early onset Alzheimer's disease patients to make involve naming a surrogate decision-maker, and genetic testing, so that potentially at-risk family members can be identified and warned.

Disclosure Issues

An important ethical issue raised in this film revolves around disclosure of the illness to others: family members, work colleagues, friends, etc. (Alzheimer's Association 2017b). In the film, Alice delays telling her Department Chair about her diagnosis until it affects her ability to work. She delays telling her spouse she's been seeing a neurologist until she has no choice; and finally, she is faced with disclosing it to her adult children, due to the genetic nature of the disease. Generally, disclosure of Alzheimer's disease has a profound effect on the patient's social location, and social circle; it can lead to various stages of social death prematurely. Alice's diagnosis means she has to lose her professional life; it doesn't mean she loses her friends and family, however, but disclosure of dementia means one is disclosing a disability, and the consequences can lead to increasing isolation as the social activities diminish. It's important to discuss the right to privacy and concealment, too; typically, the disease progression and obvious changes force most patients into wanting to explain their behaviors through disclosure. But early disclosure also helps with coping and support. There may be an ethical duty to warn in cases where patients with a genetic mutation for early onset Alzheimer's disease refuse to disclose it to their at-risk relatives, thereby preventing them from taking advantage of options such as PGD.

Privacy regarding dementia or genetic test results is another ethical issue that prompted passage of the Genetic Information Non-Discrimination Act (GINA), which you could discuss more with students, depending on the audience.

Ethical, Legal and Social Implications of Genetic Screening

The character of Anna, Alice's affected daughter uses her test results to take advantage of PGD, which is widely available in this time frame, and been for decades. But Alice's youngest daughter, Lydia, chooses not to get tested because she doesn't want to know. However, if Lydia were married with children, what would we do about her refusal to get tested, which impacts her at-risk children? For adult onset disorders, children do not need to know until they reach adulthood, a precedent established with Huntington's disease. When teaching *Still Alice*, you could cover the concepts of duty to warn in a genetic context. The legal and ethical duty to warn identifiable third parties of foreseeable, serious harm was established in *Tarasoff. V. Regents of the University of California* (1976), in which the court held that "privacy ends where the public peril begins". In Tarasoff, the failure to warn a patient's girlfriend (Tatiana Tarosoff) about her premeditated murder (the patient was Prosenjit Podar) led to a new standard about warning third parties who are not direct patients. The Tarasoff ruling is a translational health law precedent that has been widely applied to numerous clinical contexts in which the patient may be an "agent of harm" (Rosenthal and Pierce 2005; Tarasoff v. Regents 1976; Pate v. Threlkel 1995), either wittingly, or unwittingly. Thus, the duty to warn is now a standard of care in adult clinical ethics cases for both infectious disease and highly penetrant genetic diseases (Rosenthal and Pierce 2005; Molloy v. Meijer 2004; Safer v. Pack 1996). In a genetic context, the patient needs to be warned directly, and asked to participate in constructing a family tree so that at-risk relatives can be contacted and given the opportunity to be screened. Here, the ethical dilemma only presents when the patient refuses permission to contact at-risk relatives, and/or states that s/he does not wish to disclose his/her health information to family members. In this context, there are situations where it is ethically permissible, or even obligatory, to warn third parties without the patient's consent, but we would only do so with participation of a clinical ethics consultation and an ethics opinion about the nature of the harm, and whether we would consider it "imminent". Most would agree that adult children in their 20s should be warned so they can take advantage of PGD.

Conclusions

Still Alice is the neurology equivalent of *Wit* (Chap. 1)—it is a rich film that may profoundly affect the viewer due to resonance with the audience. There are many cutting-room floor decisions to make, but this film, at its heart, forces a clinical ethics discussion about flawed "bright line" approaches to capacity, autonomy and advance directives. Because *Still Alice* also deals with an autosomal dominant form of Alzheimer's disease that is 100% penetrant, it also can be the pathway to much deeper genomic discussions. We discuss many issues in this chapter, and depending upon the curriculum, there are many decisions to make in terms of which issues to highlight. Most of all, it is also a realistic clinical picture of Alzheimer's disease patients and patient encounters for a neurology audience, who need to understand these nuanced questions surrounding capacity.

Stats and Trivia

- Moore brought Alec Baldwin to the project because they worked together on the sitcom, *30 Rock*.
- During filming, Glatzer used an iPad to direct, due to his declining physical condition. Moore won the academy award for Best Actress for her portrayal of Alice in 2015, and she dedicated her Academy Award win to Glatzer, who died from ALS in March 2015.
- Ironically, both Moore and Baldwin became notable for portrayals of political figures. Moore was cast as Sarah Palin in the film *Game Change* (2012), about the 2008 election. In 2016, Baldwin became the satirical "Donald Trump" on *Saturday Night Live*, a one-time performance that became an ongoing commentary of the 45th President of the United States and his administration.

From Theatrical Poster

Director: Richard Glatzer and Wash Westmorland
Producer: James Brown, Pamela Koffler, Lex Lutzus
Screenplay: Richard Glatzer and Wash Westmorland
Based on: Still Alice by Lisa Genova
Starring: Julianne Moore Alex Baldwin, Kristen Stewart, Kate Bosworth
Music: Ilan Eshkeri
Cinematography: Denis Lenoir
Editor: Nicolas Chaudeurge
Production company: Killer Films, Lutzus-Brown, BSM Studio, Big Indie Pictures, Shriver Films
Distributor: Sony Picture Classics
Release Date: September 8, 2014 (Toronto International Film Festival), January 16, 2015.
Runtime: 101 min

References

Alzheimer's Association. Early Onset Alzheimer's Disease. 2017a. Posted to: http://www.alz.org/alzheimers_disease_early_onset.asp. Accessed 24 Feb 2018.

Alzheimer's Association. 2017b. Sharing Your Diagnosis. https://www.alz.org/i-have-alz/sharing-your-diagnosis.asp. Accessed 19 Jul 2017.

American Medical Association. 2009. Opinion 2.131—Disclosure of familial risk in genetic testing. AMA Code of Medical Ethics. *AMA Journal of Ethics* 11:683–9. http://journalofethics.ama-assn.org/2009/09/code1-0909.html.

Bailey, Ronald. 2013. FDA shuts down 23andMe: outrageously banning consumer access to personal genome information. *Reason*, November 25. http://reason.com/blog/2013/11/25/fda-shuts-down-23andme-outrageously-bann.

Baltimore, D., P. Berg, M. Botchan, D. Carroll, R.A. Charo, G. Church, J.E. Corn, G.Q. Daley, J. A. Doudna, M. Fenner, H.T. Greely, M. Jinek, G.S. Martin, E. Penhoet, J. Puck, S.H. Sternberg, J. S. Weissman, and K.R. Yamamoto. 2015. A prudent path forward for genomic engineering and germline gene modification. *Science* 348: 36–38. http://science.sciencemag.org/content/348/6230/36.

Beckham, Beverly. 2008. Despite Monster, she is 'Still Alice'. *Boston Globe*, March 16. http://archive.boston.com/news/local/articles/2008/03/16/despite_monster_she_is_still_alice/.

Beede, D., T. Julian, D. Langdon, G. McKittrick, B. Khan, and M. Doms. 2011. Women in STEM: A Gender Gap to Innovation. ESA Issue Brief #4-11. (Report). U. S. Department of Commerce, Economics and Statistics Administration. http://www.esa.doc.gov/sites/default/files/womeninstemagaptoinnovation8311.pdf.

Brandom, Russell. 2013. Body blow: How 23andMe brought down the FDA's wrath. *The Verge*, November 25. https://www.theverge.com/2013/11/25/5144928/how-23andme-brought-down-fda-wrath-personal-genetics-wojcicki.

Casey, D.A., D. Antimisiaris, and J. O'Brien. 2010. Drugs for Alzheimer's Disease: Are They Effective? *Pharmacy and Practice* 35: 208–211. https://www.ncbi.nlm.nih.gov/pmc/articles/PMC2873716/.

Cyronski, David and Sara Reardon. 2015. Chinese scientists genetically modify human embryos: Rumours of germline modification prove true—and look set to reignite an ethical debate. *Nature* (news), April 22. http://www.nature.com/news/chinese-scientists-genetically-modify-human-embryos-1.17378.

Elwood, Gregory. Julianne Moore, Kristen Stewart and Maria Shriver: 'Still Alice' home to L.A. 2014. *Uproxx*, November 13. http://uproxx.com/hitfix/julianne-moore-kristen-stewart-and-maria-shriver-bring-still-alice-home-to-la/.

English, Bella. 2016. Author Lisa Genova turns scientific fact into fiction. *Boston Globe*, May 9. http://www.bostonglobe.com/lifestyle/2016/05/08/author-lisa-genova-turns-scientific-fact-into-fiction/KiQfOrsYud9cj5O7Y3deAO/story.html?event=event25?event=event25.

Ethics Committee of the American Society for Reproductive Medicine. 2013. Use of preimplantation genetic diagnosis for serious adult onset conditions: a committee opinion. Fertil Steril 2013; 100:54–7.

Fainaru-Wada, Mark, and Steve Fainaru. 2013. *League of Denial: The NFL, Concussions, and the Battle for Truth*. New York: Crown Archetype/Crown Publishing Group.

Farhi, Paul. 2015. Maria Shriver reported on a movie about Alzheimer's for NBC. She didn't mention she's one of the film's executive producers. *Washington Post*, January 7. https://www.washingtonpost.com/lifestyle/style/maria-shriver-reported-on-a-movie-about-alzheimers-for-nbc-she-didnt-mention-shes-one-of-the-films-executive-producers/2015/01/07/deea3dde-967a-11e4-927a-4fa2638cd1b0_story.html?utm_term=.186fb6cb7b8a.

Genova, Lisa. 2009. *Still Alice*. New York: Simon and Schuster.

Handyside, A.H., E.H. Kontogianni, K. Hardy, and R.M. Winston. 1990. Pregnancies from biopsied human preimplantation embryos sexed by Y-specific DNA amplification. *Nature* 344:768–770. https://www.ncbi.nlm.nih.gov/pubmed/2330030.

Kirk, Michael, Jim Gilmore, and Mike Wiser. 2013. *League of Denial*. Frontline. Public Broadcasting System, released October 8. http://www.pbs.org/wgbh/frontline/film/league-of-denial/.

Kolata, Gina. 2015. Chinese scientists edit genes of human embryos, raising concerns. *New York Times*, April 23. http://www.nytimes.com/2015/04/24/health/chinese-scientists-edit-genes-of-human-embryos-raising-concerns.html?_r=0.

Kushner, Tony. 2013. *Angels In America*. Revised and Completed Edition. New York: Theatre Communications Group.

Lanphier, E., F. Urnov, S.E. Haecker, M. Werner, and J. Smolenski. 2015. Don't edit the human germline. *Nature* 519: 410–11. http://www.nature.com/news/don-t-edit-the-human-germ-line-1.17111.

Learner, Sue. 2016. Still Alice helped trigger 'global conversation about Alzheimer's' giving people with dementia a voice and a face. *Homecare*, July 7. https://www.homecare.co.uk/news/article.cfm/id/1577032/still-alice-trigger-global-conversation-alzheimers-dementia.

LeMire, Christy. 2014. Still Alice. RogerEbert.com, December 5. http://www.rogerebert.com/reviews/still-alice-2014.

Liang, P. Xu, X. Zhang, C. Ding, R. Huang, Z. Zhang, J. Lv, X. Xie, Y. Chen, Y. Li, Y. Sun, Y. Bai, Z. Songyang, W. Ma, C. Zhou, J. Huang. 2015. CRISPR/Cas9-mediated gene editing in human tripronuclear zygotes. *Protein and Cell* 6: 363–372. http://link.springer.com/article/10.1007%2Fs13238-015-0153-5.

Madigan, Nick. 2002. Charlton Heston reveals disorder that may be Alzheimer's Disease. *New York Times*, August 10. http://www.nytimes.com/2002/08/10/us/charlton-heston-reveals-disorder-that-may-be-alzheimer-s-disease.html.

Maron, Dina. 2015. Improving humans with customized genes sparks debate among scientists. *Scientific American*, December 3. https://aws.scientificamerican.com/article/improving-humans-with-customized-genes-sparks-debate-among-scientists1/.

Malloy v Meier, 629 NW2d 711 (Minn 2004).

Misra, J., J.H. Lundquist, and S. Agiomavritis. 2011. The Ivory Ceiling of Service Work. *The American Association of University Professors (AAUP) Report*. January 1. https://www.aaup.org/article/ivory-ceiling-service-work#.WTBDGxT-yOo.

Mullin, Emily. 2017. FDA opens genetic floodgates with 23andMe decision. *MIT Technology Review*, April 6. https://www.technologyreview.com/s/604109/fda-opens-genetic-floodgates-with-23andme-decision/.

National Human Genome Research Institute. 2008. The Genetic Nondiscrimination Act. https://www.genome.gov/24519851/. Accessed 24 Feb 2018.

Pate v Threlkel, 661 So 2d 278 (Fla 1995).

Pollack, Andrew. 2015. F.D.A. Reverses Course on 23andMe DNA Test in Move to Ease Restrictions. *New York Times*, February 19. https://www.nytimes.com/2015/02/20/business/fda-eases-access-to-dna-tests-of-rare-disorders.html.

President's Council on Bioethics. 2004. Reproduction and Responsibility: The Regulation of New Biotechnologies (Report). https://bioethicsarchive.georgetown.edu/pcbe/reports/reproductionandresponsibility/index.html.

Rosenthal, M. Sara. and Pierce, Heather H. 2005. Inherited medullary thyroid cancer and the Duty to Warn: Revisiting Pate v. Threlkel. Thyroid 15:140–45.

Rosenthal, M. Sara. 2015. Preimplantation Genetic Diagnosis: Safe Sex in a Gattacan World. *Endocrine Ethics Blog*, July 24. http://endocrineethicsblog.org/preimplantation-genetic-diagnosis-safe-sex-in-a-gattacan-world/. Accessed 24 Feb 2018.

Safer v Estate of T. Pack, 677 A 2d 1188 (NJ Supp 1996).

Seymour, Tom. 2015. Still Alice is 'shockingly accurate' – people living with dementia give their verdict. *The Guardian*, February 10. https://www.theguardian.com/film/2015/feb/10/still-alice-alzheimers-accurate-dementia-sufferers-verdict.

Smith, Nigel M. 2015. Robin Williams' widow: 'It was not depression' that killed him. *The Guardian*, November 3. https://www.theguardian.com/film/2015/nov/03/robin-williams-disintegrating-before-suicide-widow-says.

Tarasoff v Regents of University of Calif, 551 P2d 334 (1976).

U.S. Equal Employment Opportunity Commission. 2008. The Genetic Nondiscrimination Act. https://www.eeoc.gov/laws/statutes/gina.cfm. Accessed 24 Feb 2018.

United States Congress. 2008. *The Genetic Non-Discrimination Act of 2008*. Public Law 110-233. U.S. Government Printing Office. https://www.gpo.gov/fdsys/pkg/PLAW-110publ233/html/PLAW-110publ233.htm.

Yin, Steph. 2015. Germline engineering could lead to designer babies and superstrength. *Motherboard*, April 15. http://motherboard.vice.com/read/germline-engineering.

The Principle of Beneficence is a specific obligation to ensure that our intention is to *improve* patient well-being by maximizing benefits and minimizing harms. Beneficence goals of care must, therefore, take into account potential psychosocial benefits and harms as much as medical benefits and harms. When healthcare providers encounter an ethical dilemma surrounding beneficence, it is usually the *healthcare provider's* dilemma; patients may not be aware of the dilemma themselves, but healthcare providers may agonize over a plan of care that meets the beneficence standard, so that they do not "cross a line" into what they may perceive as a violation of the Principle of Non-maleficence—the specific duty not to intentionally cause harm through an act of commission or act of omission (neglect), which may also include an ethical duty to warn (See Chap. 7). The Principle of Beneficence also guides the "Best Interest" standard that is used by surrogate decision-makers for adults (see Part II), and "Best Interests of the Child" in a pediatric ethics framework. Thus, the Principle of Beneficence is an autonomy-limiting principle in cases where a surrogate decision-maker may be making medical decisions that lead to more harm than good for the patient involved.

Each film in this section is a dramatization of an actual case. In *The Elephant Man*, the nineteenth-century case of Joseph Merrick (a.k.a. John Merrick) challenges us to consider whether his physician, Frederick Treves, is a "good man or a bad man"—a question Treves asks of himself, when he wonders whether he is helping or exploiting his patient, who was given hospice at the London Hospital. In *Lorenzo's Oil*, we revisit the 1980s case of Lorenzo Odone, whose parents do not accept his terminal diagnosis and insist on experimental treatments that meet the standard of beneficence from a research ethics framework but may not meet that standard from a clinical ethics framework or a pediatric ethics framework. Finally, in Awakenings, the perfect laboratory to study beneficence, we revisit the work of neurologist Oliver Sacks and the early clinical trials of L-DOPA conducted in 1969 with a group of patients with strange neurological symptoms that leave them motionless after being struck with "sleepy sickness" in a pandemic (1916–1928). Sacks experiences a profound beneficence dilemma when a "miracle drug" is short-lived: at first, it offers great benefit, but then after a "honeymoon window", it causes more harm than benefit.

Other Core Bioethics Principles Involved

Although the core bioethics principle dominating in this section is the *Principle of Beneficence,* the conjoined Principle of Non-maleficence, which is the specific obligation not to knowingly cause harm, or render a harmful therapy, is always a balancing act that helps keep the main goal of beneficent care plans in check. A greater balance of benefits than harms is sometimes unclear, as in all three of the films in this section. Upholding The Principle of Respect for Persons is why we must pay attention to Beneficence; when we are protecting the vulnerable or non-autonomous patients (as in Joseph Merrick, Lorenzo, and Leonard, in *Awakenings*), we may need to limit autonomy when we feel what the patients or surrogates want is not in their best interests.

Finally, the Principle of Justice in this section weaves into discussions about clinical trial candidacy and selection (in Lorenzo's Oil and Awakenings) in the absence of any standard therapy or treatment, costs of trials and treatments, and even in Merrick's case, the cost of hospice for a disabled patient. Specific ethical issues are explored in each chapter in this section.

"Am I a Good Man, or a Bad Man"? The Elephant Man (1980)

8

David Lynch's *The Elephant Man* is the story of a patient, Joseph Merrick (1862–1890), who suffered from an extreme and rare deformity now understood to be caused by Proteus Syndrome, characterized by atypical growth of the bones, skin, head and a variety of other symptoms. This condition was first identified by Michael Cohen Jr. in 1979 (Cohen 1986; Tibbles and Cohen 1986; Cohen 1988a, b), inspired by a resurgence of medical interest in the Merrick case in the 1970s. The case history of Joseph Merrick was fully documented by Merrick's physician, Sir Frederick Treves, in his memoir The Elephant Man and Other Reminiscences (1923). Ironically, Tod Robbins' short story "Spurs" was also published in 1923; it was the source material for Tod Browning's *Freaks* (1932), which was written by Robbins (Wilson 2001), and featuring the Siamese twins, the Hilton Sisters (Wilson 2001; Darke 1994) often compared to David Lynch's *The Elephant* Man (Darke 1994). Treves' memoir is published during the aftermath of World War I, and resonated with a generation that was disfigured or maimed in the war.

A later chronicle of Merrick's life was published in 1971 by anthropologist, Ashley Montagu, entitled The Elephant Man: A Study in Human Dignity. Montagu was Jewish, and experienced anti-Semitism throughout his life; his body of work was in the area of race, humanism and dignity, and his interest in Merrick helps to explain this film's producer: Mel Brooks (produced under Brooksfilms). Brooks also stated that Merrick's story resonated with his Jewish upbringing, and feelings of being persecuted as an outsider (Paramount 2001). Yet fascination with monstrosity was not a new theme for Mel Brooks, whose film, *Young Frankenstein* (1974), explored similar themes of freak and medicine. Brooks was impressed with David Lynch's first film, *Eraserhead* (1977). Brooks offered Lynch the opportunity to direct *The Elephant Man*, and arranged to have his own name omitted from the credits to ensure that no one expected farce or comedy. Lynch recalls in an interview (Huddelston 2008), that upon completion of *Eraserhead* in 1977:

© Springer International Publishing AG, part of Springer Nature 2018
M. S. Rosenthal, *Clinical Ethics on Film*,
https://doi.org/10.1007/978-3-319-90374-3_8

> I wrote a script called "Ronnie Rocket", but I couldn't anything going. I met a man named Stuart Cornfeld, who worked for Mel Brooks and had loved "Eraserhead". One day, just on a feeling, I said, "We're not getting anywhere with 'Ronnie Rocket': are there any other scripts that I might direct? And he said "There are four scripts. Come to Nibblers [a restaurant] and have lunch, and I'll tell you. "The first thing he said was "The Elephant Man". And an explosion went off in my brain. Very strange. I said immediately, "that's it. That's what I want to do."

Montagu's book incorporated much of Treves' original memoir, but also added other accounts of Merrick, based on additional sources, and corrects facts that Treves has misconstrued. Interest in Merrick ensued after the Montagu book was published. A stage play and a screenplay entitled "The Elephant Man" were being independently generated around the same time in the late 1970s. (Apparently, Brooks was unable to acquire the rights to the stage play, and began his own film adaptation from the same source material). Additionally, a novel entitled The Elephant Man (1980) by Christine Sparks, was a novelization of the Lynch film. Ultimately, the film is an entirely different adaptation of Treves' and Montagu's accounts than Bernard Pomerance's 1977 stage play, which debuted on Broadway in 1979 and won a Tony Award (David Bowie was cast in the lead in some productions). In fact, the film was already in full production and being shot in London before the play came to Broadway, but the play debuted before the film was released. Victor Canby noted in his October 3, 1980 review of the film's debut that: "David Lynch's haunting new film [is] not to be confused with the current Broadway play of the same title, though both are based on the life of the same unfortunate John Merrick, the so-called Elephant Man, and both, I assume, make use of the same source materials" (Canby 1980).

Between the debut of the play and the Lynch film, however, another work about Merrick is released in the scholarly world. In the Spring of 1980, English scholars Michael Howell and Peter Ford published The True History of the Elephant Man, which presented new information about the Merrick case, including his given name of Joseph and not, John, Merrick. Additionally, Howell and Ford refute much of what is in Treves' account. The Howell and Ford presentation of the case was not the source material Lynch used, however, for his film.

In Lynch's film treatment, *The Elephant Man* doesn't just tell the story of Merrick, but the story of public response to disfigurement and disability. The story is as much about our fascination with "freaks" as it is about the experience of being persecuted from the "freak's" perspective. Ultimately released in 1980—a full decade before the *American With Disabilities Act* of 1990—*The Elephant Man* is perhaps more about the audience than it is about the man. And this was likely Lynch's intention.

As for the film's title and Merrick's "nickname": Merrick himself attributed his condition to a real incident that occurred during his mother's pregnancy. She had been knocked over by an elephant while she was pregnant. In Victorian England, the theory of "maternal impression" (Kochanek 1997; Darke 1994) was what was used to explain birth defects or congenital disorders. Merrick was convinced that this incident explained his deformed appearance resembling an elephant, and why he was called the "Elephant Man." Lynch makes full use of this maternal impression tale by stretching it further: he begins the film with a scene that strongly suggests Merrick's mother was actually raped by an elephant. For the theatre audience of 1980, however, the connection between prenatal events and deformity is made. Some film scholars suggest Lynch also works with Christian perceptions of sin (Johnson 2003), suggesting Merrick's deformity was somehow a punishment arising from coupling of beast and human (Darke 1994). When on "display" in the medical world, Merrick produces a picture of his beautiful mother for Treves and his wife, and remarks that he must have been a "great disappointment" to his mother. I suggest this played on modern fears of women who were delaying childbirth in the wake of feminism—a delay with biological consequences of increasing their chances of having children with fetal or chromosomal abnormalities with advancing maternal age.

Synopsis

Released in 1980, and nominated for eight Academy Awards in 1981, including Best Picture, *The Elephant Man* was adapted for the screen by David Lynch, Christopher De Vore, and Eric Bergren from the Treves and Montagu books about Merrick. The film begins with the mythical origins of Merrick's unfortunate beginnings: that his mother was knocked down by an elephant, which is supposed to explain his unusual deformities—large tumors on his skin, misshapen bones and enlarged skull—an overall appearance that is "elephant-like". When the actual story begins, it is faithful to its source materials. A surgeon at the London Hospital (Treves, played by Anthony Hopkins) discovers John Merrick (played by John Hurt) in a freak show. He arranges for a private viewing of Merrick with Merrick's manager, who mistreats Merrick. Treves is clearly fascinated and moved by Merrick, although we don't see what Merrick actually looks like yet—only Treves' expression of intrigue and emotion as he tears up upon his viewing of Merrick. Treves' then pays the manager to bring him to the London Hospital so that he can examine him properly. There, Treves presents Merrick to his colleagues at The Pathological Society, as a grand rounds lecture. Essentially, Treves puts Merrick on display once more—only this time—for a medical audience. He states in his presentation that: "At no time, have I met with such a perverted or degraded version of a human being as this man. I wish to draw your attention to the insidious conditions affecting this patient." He then goes on to describe the deformities in detail, commenting that, "as an interesting side note" Merrick's genitals are unaffected and completely normal.

After Merrick returns to his home at the freak show, he suffers severe beatings from his manager, and is brought back to Treves for treatment by one of the freak show workers. Treves presents Merrick as a charity case to the hospital, and begins to lobby to provide permanent hospice to Merrick, where he can be made comfortable. During this process, Treves is alarmed to discover that Merrick is not an "imbecile" as previously described by Merrick's manager, and which appears to be the case due to Merrick's silence. When Merrick finally speaks, Treves finds that Merrick is highly intelligent, articulate, and culturally refined, which makes his physical deformity all the more tragic. When Merrick is revealed to be a fully autonomous man, his suffering is revealed to be all the more profound. Thus, much of the disability becomes Merrick's own awareness of his deformities. Ultimately, Merrick is granted permanent residence at the hospital, and is introduced to London's upper class society, who, we presume, are interested in meeting Merrick because they are actually more intrigued with his freakishness than they are with his personality. Treves begins to question whether he has done more harm than good for this patient in the final analysis, and compares himself to the freak show exhibitor. He asks his wife, in a moment of epiphany: "Am I a good man, or a bad man?"

The plot thickens as Merrick is "stolen" back from the hospital by his original freak show manager, who takes him out of England and displays him across Europe. Eventually, Merrick is helped by his fellow freak exhibits to buy passage back to London, and makes his way back to the hospital. When travelling in public, Merrick must cover his head with a hood (which has two peepholes for his eyes) to avoid shocking people. He is harassed on his journey back to the hospital, and chased down and cornered by a horrified public. At this point, the famous line escapes from his tortured soul: "I am NOT an animal. I am a human being."

Lynch shot the film in black and white, which helped to frame it as nineteenth century story. Additionally, for technical reasons, black and white film helped to enhance the makeup, and made it look more realistic (Paramount 2001).

When I screen and teach *The Elephant Man* for my students, I frame it as a case surrounding the bioethics principle of Beneficence, which obligates the healthcare provider to maximize benefits and minimize harms. However, in so doing, the principle of non-maleficence (the specific obligation not to *intentionally* do harm), may be breached. Ultimately, I suggest to my students that beneficence and maleficence (a deliberate, harmful act) are frequently conjoined. Frederick Treves begins to realize this in the Lynch film, and hence, his question to himself, in a moment of intellectual honesty, unveils the terrible truth: he believes he has done some harm, and he is not at all certain that his intentions were "benevolent" or innocent. He has indeed made Merrick a "curiosity all over again" by reframing his freakishness as a medical oddity. Worse, he has perhaps done himself the greatest professional service at Merrick's expense. "What was it all for?" He wonders. "Am I a good man, or a bad man?" Treves' wife tells him what he wants to hear, of course. "John Merrick is happier and more fulfilled now than he's ever been in his entire life, and it's completely due to you." But deep down, Treves knows the truth: Merrick is "his" freak who has made his career. As Darke (1994) observes: "What Merrick has done for Treves far outweighs what Treves did for Merrick. Merrick has made Treves a celebrity, a rich man… he also helps Treves get a knighthood."

We cannot fully accuse Treves of exploitation, however, because as watchers of the film, we are all complicit in our fascination with Merrick as a physiologic anomaly. In the scene in which Treves introduces Merrick as a lecture topic, Lynch reveals Merrick only in shadow behind a curtain; the film watcher becomes part of the Grand Rounds audience, and we, too, try to strain to see Merrick behind the curtain (Friedman 2000). Thus, the clinical ethics themes in *The Elephant Man* are informed by a fuller presentation of the sociological, medical and historical contexts. This chapter will also explore why the nineteenth century case of Joseph Merrick finds a comfortable home, and fascinated audience, in the late twentieth century.

Why 1980? The Social Location of David Lynch's *The Elephant Man*

What compelled American artists from different media to bring the British Joseph Merrick's story to life around 1979/1980? Scholars of both the film, and Merrick's life, point to Montagu's popular book as the impetus for a revived interest in Merrick (Ablon 1995); indeed, Montagu's book captured the attention of individuals in a variety of academic, medical and artistic disciplines (Ablon 1995). In a Preface to the 1979 edition of his book, Montagu (1979) writes:

> The response to [my 1971] book has been quite astonishing. To my knowledge, there exist at least half-a-dozen movie scripts and at least eight plays by different authors. Seven of these plays have been produced. As a report in Variety (7 March 1979) put it, 'A herd of Elephant men' is proliferating on U.S. stages" (p.xiv).

The popularity of a particular book, however, is also reflective of what is bubbling beneath the societal surface. One can speculate that this was a period in which Americans were, once more, beginning to deal with disfigurement arising from war injuries. This time the war was very different: the Vietnam War, which had only formally ended in 1975. As a result of the Vietnam War, 58,148 Americans were killed in combat; over 300,000 veterans were wounded, and many had injuries that deformed the body in some way. *Coming Home* (1978) boldly looked at the emotional and physical wounds of the Vietnam War, which juxtapose spinal cord injuries against the hidden wounds of Post-Traumatic Stress Disorder (PTSD). That film also starred Jane Fonda, a notable critic of the Vietnam War. (She was labeled "Hanoi Jane".)

This period in social history, in the aftermath of the Civil Rights movement of the previous decade, saw the birth of the disability rights movement, and introductory laws that helped pave the way for the 1990 *Americans with Disabilities Act*. Disability rights advocates shifted public discourse from pitying the disabled to recognizing them as competent individuals who could make a societal contribution. Legislation began to be enacted that removed societal barriers and created equal opportunity. One major piece of legislation was *The Rehabilitation Act* of 1973, which provided for the establishment a number of programs that enabled independent living with appropriate social supports. This law also banned discrimination against people with disabilities by recipients of federal financial assistance.

This law was followed by *The Education of All Handicapped Children Act* of 1974, which mandated equal education opportunities for disabled children.

A more subverted social and bioethical topic that was challenging American thinking was the legalization of abortion (Roe v Wade 1973) and emergence of prenatal diagnosis as part of mainstream obstetrics. These medical procedures combined to allow women to presumably have more control over their social and reproductive destinies and decide whether they wished to prevent a harmed life from coming into the world. These medical procedures also offered more choices to women who wished to delay traditional social roles and delay reproduction while increasing the odds of having a child with fetal anomalies. Other films released in 1980 reflected a definitive change in how women were thinking about traditional roles (and by extension, gave fuel to the feminist adage "biology is not destiny"). The comedy *9–5* was a social commentary about economic and social inequities in the pink-collar workforce. The drama *Coal Miner's Daughter* was the true story of Loretta Lynn, who broke out of her traditional social role in spite of almost being enslaved by it. The Loretta Lynn story served as a "slavery tale" about women overcoming extremely oppressive roles (Lynn was married at 13 with four children by age 20). *Private Benjamin* was a liberating comedy about a wealthy, privileged woman who joins the army after becoming a widow and realizing she has no skills. While none of these 1980s "chick flicks" were actually about terminating a pregnancy, they were about a major shift in women's thinking about how they wanted to live. Becoming architects of their own destinies, however, could not be made possible without control over their reproductive destinies through reproductive technologies.

More broadly (with uncomfortable eugenics overtones), these technologies allowed society to question (perhaps as they did in Treves' time), who, as Harvard biologist and social scientist, Ruth Hubbard, phrased it (1997): "should and should not inhabit the world". Although amniocentesis was available by the late 1960s (Resta 1997), it was reserved for only high-risk pregnancies in which a fetal abnormality was strongly suspected, and it had limitations on what it could find. By the late 1970s, as women began to have babies at older ages, amniocentesis emerged as a routine procedure. By 1977, coinciding with the early production of *The Elephant Man*, ultrasound equipment became commercially available, and began to be routinely offered to women (Resta 1997). In 1978, the ability to produce an embryo in a test tube was achieved by Patrick Steptoe and Robert Edwards (now known as in vitro fertilization), a success story announced with the birth of Louise Brown, known as the world's first "test tube" baby (Evening News 1978; Dow 2017).

The release of *The Elephant Man* coincides with the dawning of a long trend in daytime television, known as the "tabloid talk show" which served as the modern day "freak shows". This trend peaked in the late 1980s and early 1990s, but began to unveil in the 1970s. Credited with pioneering this genre was Phil Donahue, whose talk show *Donahue* aired from 1970–1996. *Donahue* was the first to begin featuring "taboo" topics or "freak" guests who were either considered to be social "freaks" or physical freaks of nature. In addition to having on his Dayton,

Ohio-based show guests who were atheists and homosexuals, in 1978, *Donahue* had as guests Louise Brown and her parents, for example. That same year, he also devoted an entire show to conjoined twins—one set who were separated and one set who remained conjoined. The conjoined twins episode long resonated as one of the early signs that American culture was still intrigued with "freak shows". In fact, Leslie Fiedler, who, in 1978, was touring his academic book <u>Freaks: Myths, and Images of the Secret Self</u> writes in 1996 (Fiedler 1996):

> When, nearly two decades ago, I published my pioneering study of the way in which self-styled "normals" perceived certain of their physically anomalous fellow humans tra-ditionally called "freaks," scarcely any of my academic colleagues responded positively....
> When I first began the book about freaks, they were starting to dominate the postprint media. The story of the "Elephant Man" for instance, was attracting record audiences in movie houses and on television...Worst of all was my experience with Phil Donahue, when I found myself confronted (with no previous warning on either side) by a pair of conjoined twins and cast somehow as their heartless exploiter.

Other films released in 1980 that worked with themes of the "outsider" or social "freakishness" included *Ordinary People* (in which the main character, a teen who attempted suicide, is actually called "Freak" by his peers); *Fame* (which had as one of its main characters, a homosexual student struggling with coming out); the remake of *The Jazz Singer* (with Neil Diamond), about a man raised as an orthodox Jew wanting to be a mainstream singer; *Melvin and Howard* (with a focus on Howard Hughes); and *Heaven's Gate,* which dealt with themes of foreigners and outsiders (the film was also notorious for its box office failure).

Deformity and Disability: The History and Sociology of Medicine Context

There are four history of medicine contexts applicable to this analysis: (1) the historical context of British medicine in the late nineteenth century—Merrick's and Treves' actual timeframe (Merrick lived 1862–1890); (2) the historical context in which Treves' publishes his memoir (1923); (3) the historical context of Montagu's scholarship of Merrick's life (his book is published in 1971); (4) and finally, the historical context in which the Lynch film is produced and released (1978–1980). Ultimately, it is only in the latter timeframe in which the disease that afflicted Merrick (Proteus syndrome) is even discovered and named. It would be a few more years, however, before Merrick would be properly diagnosed with Proteus syn-drome (1986)—over 100 years since his first encounter with Treves. For the pur-pose of this discussion, the focus of the history of medicine analysis here is on deformity and disability.

Mid-late 1800s

Sir Frederick Treves and Joseph Merrick were born in the same generation. Treves (1853–1923) sees Merrick in 1884 in a freak show exhibit across from the London Hospital; he gives Merrick hospice from 1886–1890, when Merrick dies (Medical News 1890). Merrick was born in 1862. In Merrick's and Treves' own lifetime, the freak show was alive and well, which provided those with physical defects a forum in which to earn a living as an "exhibit" or "performer". Historians and disability scholars note that many individuals with deformities were dependent on the freak shows for their livelihoods, and were generally treated with respect by their managers and exhibitors. Merrick, like many of his "freak" peers, actually sought management and volunteered to be exhibited (Durbach 2009).

In 1859, a significant event occurs in science and medicine. The theory of evolution proposed by Charles Darwin is published in his work <u>On the Origin of Species</u>. Merrick's birth in 1862 essentially coincides with this groundbreaking text. Although Darwin's theory generates cultural and social anxieties about evolution (Durbach 2009; Toulmin 2009; Aguirre 2010) because it challenges religious and mythical explanations for human existence, the theory of evolution and natural selection reframes physiological anomalies as "scientific" rather than "demonic". This cultivates both a scientification and medicalization of "freakishness" for the lay public that did not previously exist. Thus, public fascination with a "freak of nature" replaces fearing someone with a birth defect, the disabled or disfigured. Within Merrick's lifetime, public fascination is slowly supplanted by public sympathy. This is a very different experience than earlier freak predecessors. For example, when conjoined twins, Eng and Cheng were born in 1811 in Thailand (formerly Siam, hence the phrase "Siamese Twins"), they were considered to be an omen of the end of the world. Although they eventually found fame through freak show touring under the management of a Scottish merchant, and later, through P. T. Barnum who they were with until 1839, they died as objects of exploitation and fascination rather than sympathy. After their deaths, an autopsy performed by the College of Physicians of Philadelphia determined they could have been successfully separated; it was a medical option that was never offered to Eng and Chang during their lives—presumably because their value as an exhibit was at stake. Merrick's experience, in a post-Darwinian world, is different; Merrick is rescued by the medical world and granted hospice. Public sympathy for Merrick builds as does disdain over the exploitation of disfigured or disabled individuals in freak show exhibits. Historians note that the freak shows in London begin to rapidly decline in popularity around the 1860s, and by the late 1800s, are no longer a mainstream activity, but relegated to a smaller "side show" which never ultimately disappeared until the early 1970s (Garland Thompson 1996).

In the medical world, fascination with physiological anomalies had been a long tradition that predated Darwin. Specimens for study and examination had begun to be collected and shown as closed exhibits only to those in the medical or scientific fields. One of the most notable collectors of unusual specimens was John Hunter, who was a Fellow of the Royal College of Physicians and Surgeons. His collection of specimens dates back to 1783; his specimens were purchased by the British government in 1785, and presented to the Royal College as a museum of pathological specimens. In Philadelphia, the Mutter Museum is established in 1858 by the College of Physicians of Philadelphia as a bequeathed collection of medical specimens and artifacts by one of its members, Thomas Dent Mutter (Kochanek 1997; Mutter Museum 2018). Of note, this museum acquired the plastic cast of Eng and Chang in 1874, when the College of Physicians of Philadelphia performed their autopsies. What begins to change in post-Darwinian medical culture is to use the medical journal as the proper forum in which to share unusual cases, and provide more rigorous analysis. Only a handful of such journals existed at this time, including the *Lancet,* which debuted in 1840. Although medical case reports appeared in medical journals prior to Darwin, they become regular features by the 1860s, and developed a lexicon that was more professional. As Kochanek (1997) observes: "Case histories are an effect of medicine's need to explore deformity empirically and of the professional need to distance medical looking from sideshow voyeurism." By the time Treves discovers Merrick in the 1880s, illustrated case reports are the norm, and the language describing the cases echoes phrases used by Darwin. For example, in 1865, "A Remarkable Case of Double Monstrosity" appears in the *Lancet* as an illustrated case, describing a man with two penises (Kochanek 1997). The author observes the penises' "equal adaptation to the general functions of the bearer" (Kochanek 1997; The Lancet 1866). In a Darwinian context, evolutionary success is all about function and usefulness to the species rather than aesthetics. In Treves' presentation of Merrick, he echoes this trend when he comments that Merrick's "genitals" are unaffected, making it clear that Merrick's deformities do not interfere with his capacity to reproduce.

In 1882, the condition of neurofibromatosis is described by Frederich von Recklinghausen (Carswell 1982), which would later be ascribed the name "Elephant Man Syndrome" (Wilkie and Rabson 1979) or "Elephant Man's disease" (Legendre et al. 2011) because it was thought (wrongly) to be the mysterious condition that Merrick suffered from.

In 1884, Merrick's case is reported in the *British Medical Journal* (Treves 1884); in 1885, Treves labels Merrick as suffering from a congenital deformity (Treves 1885a, b), and in 1888, Henry Radcliffe Crocker, a dermatologist proposed that Merrick may be suffering from two skin diseases, dermatolysis and pachdermoticele; he describes Merrick in an 1888 textbook, Diseases of the Skin: their Description, Pathology, Diagnosis and Treatment (Howell and Ford 1980).

The association with Merrick and neurofibromatosis begins in 1909, when F. Parkes Weber suggests it (Weber 1909), and then reasserts this in 1930 (see further). Speculation about Merrick's condition would persist until the mid-1980s.

An unintended consequence of Darwinism was the dawning of "social Darwinism", also known as the eugenics movement (eugenics literally means "good genes"). Ultimately, eugenics would not be in full bloom until the early twentieth century, around the time Treves publishes his memoir in 1923, but the seeds begin in the post-Darwin nineteenth century. Eugenists used medical knowledge of heredity and Darwinian theory to launch ideas about social engineering and social breeding. Ultimately, eugenics as a movement took strong hold in the early twentieth century, but was beginning to swirl around deformity, disability, and the underclass. British eugenics began in the 1860s with publication on the topic by Sir Francis Galt (Darwin's cousin), who was an animal breeder. He wrote: "What Nature does blindly, slowly, and ruthlessly, man may do providently, quickly, and kindly" (Mackenzie 1975). The first Eugenics Laboratory was established in London in 1907 (Mackenzie 1975; Elof 2018).

1920s

When Treves publishes his memoir <u>The Elephant man and Other Reminiscences</u> in 1923, he is living in post-World I Europe, which was struggling with a large population of disfigured and disabled veterans of approximately 21 million (Gerber 2003). The most notorious injuries in Europe, and the injuries which most resonated with the case of Merrick, were facial injuries. Men came home so horribly disfigured (half a face may be missing, for example), the only response was to call upon artisans to craft special masks. At the 3rd London General Hospital, a facility known by veterans as the "tin noses shop" was actually the "Masks for Facial Disfigurement Department". According to Alexander (2007), this department "represented one of the many acts of desperate improvisation borne of the Great War, which had overwhelmed all conventional strategies for dealing with trauma to body, mind and soul." London surgeon, Sir Harold Gillies, who specialized in facial reconstruction, began working with artists who could restore a mutilated man's original face by creating prosthetic masks. Another major figure in this department was Francis Derwent Wood, an artist who had been assigned to the 3rd London General Hospital when the war broke out. He saw that he could be useful as an artist in helping men recover from facial wounds. He worked with metallic masks that were lightweight and more permanent than the rubber prosthetics, and custom fit them for each veteran. Mask-making for facial wounds began around 1916, and was discussed in the *Lancet* in 1917. A similar mask studio opened in Paris in 1917, known as the Studio for Portrait Masks, which was administered by the American Red Cross. Alexander (2007) notes:

> Those patients who could be successfully treated were, after lengthy convalescence, sent on their way; the less fortunate remained in hospitals and convalescent units nursing the broken faces with which they were unprepared to confront the world–or with which the world was unprepared to confront them. In Sidcup, England, the town that was home to [a] special facial hospital, some park benches were painted blue; a code that warned townspeople that any man sitting on one would be distressful to view. A more upsetting encounter, however, was often between the disfigured man and his own image. Mirrors were banned in most wards, and men who somehow managed an illicit peek had been known to collapse in shock.

Ultimately, because of the painstaking work involved, there were not enough resources to make masks for the thousands of veterans with facial wounds who needed them, but mask-making continued until the early 1920s. Men were walking around with such masks all over Europe and some in the U.S. For those who couldn't get masks, some countries created special residences for them. In France, the "Union of the Facially Wounded" was created, for example. In the U.S., because it had entered the war later, and had fewer men fighting, facial disfigurement was not as prominent. Compared to the thousands of European men with these wounds, there were about 200–300 American men with such wounds. There was also a concerted effort to deal with the psychosocial trauma of war by providing "inspirational stories" (Gerber 2003) by persons with disabilities who overcame their challenges, or found a way to live with them.

It was amidst this context and virtual armies of shattered bodies that Treves publishes his memoir, in 1923, and then dies that same year. The story of Merrick's plight especially resonated in a post-war Europe and U.S., which were just beginning to deal with en masse disfigurement and disability as a culture. It is also in the 1920s when Lon Chaney resonates on film with those who were wounded in the war and reproduces the "disfigured or deformed male body to be visualized on screen" (Randell 2003).

However, Treves memoir begins to fade from the cultural memory, but continues to fascinate and resonate within the medical community.

By 1930, F. Parkes Weber hypothesizes that Merrick had neurofibromatosis in the *Quarterly Journal of Medicine* (Weber 1930). Thereafter, Merrick is wrongly associated with this disease until the mid-1980s, when he is appropriately diagnosed with Proteus Syndrome. However, his association with neurofibromatosis generates a renewed interest in this rare genetic disease, and sufferers become unnecessarily fearful that they will wind up like the "Elephant Man".

By this time, the eugenics movement is at its peak, and begins to capture the imagination of a particularly embittered European World War I veteran: Adolf Hitler. He uses eugenics theory to develop social cleansing policies as he rises to power through the 1920s and 1930s. Such policies favor exterminating on the basis of weakness, disability, and, of course, the Jewish race. This would generally become known as the Holocaust, but the disabled, deformed or disfigured were annihilated, too. Two Jews who are living in a post-Holocaust world find their own experiences of persecution resonating in Merrick's tale: Ashley Montagu and Mel Brooks.

1971–1986

When Ashley Montagu publishes his 1971 book entitled <u>The Elephant Man: A Study in Human Dignity</u> (1971) about Merrick, there is renewed interest in Merrick on a number of artistic and social fronts, as previously discussed. In the medical world, Merrick's story resonates in several fields. On the genetics front, interest in both neurofibromatosis, and postulating other genetic diseases as Merrick's plight generated several investigations and medical journal papers. For example, Wilkie and Rabson (1979) discuss Merrick's case in the journal, *Aesthetic Plastic Surgery* (based on a presentation in 1973 to the American Society for Aesthetic Plastic

Surgery) and propose a new term "The Elephant Man Syndrome" be used for "those so ugly by certain congenital conditions that they are unable to function happily" (Wilkie and Rabson 1979). These authors go on to make a plea that plastic surgeons apply their skills in humanitarian efforts to help these patients. Independent of this, a new genetic disorder is described in 1979 known as Proteus Syndrome (Cohen and Hayden 1979; Biesecker et al. 1999).

After the film's release in 1980, it led to a greater interest in neurofibromatosis. For example, a 1982 *Journal of the American Medical Association* article noted: "As communicated by a Broadway play, a motion picture, and recent medical and popular writings, Merrick's tormented life has unquestionably contributed significantly toward greater knowledge of the progressive, familial disorder neurofibromatosis" (Carswell 1982). A 2011 search for papers on neurofibromatosis by Legendre et al. (2011) found that between 1909 and 1986, the 26 papers published about this extremely rare disorder all used the word "elephant man". Several of these papers published illustrations of Merrick, ultimately "making him a curiosity all over again" as Treves states in the film.

Beyond 1980, intense interest in the film persisted, and led to researching a more definitive diagnosis, or challenging the presumed diagnosis of neurofibromatosis. New technologies also make Merrick a posthumous research subject. For example, it was postulated by Benjamin Felson, a professor of radiology, that Merrick suffered from three different bone diseases that contributed to his deformity. These conclusions were reached after the first x-rays were taken of Merrick's body (Carswell 1982).

By 1986, Canadian geneticists Tibbles and Cohen were able to demonstrate that Merrick did not have neurofibromatosis, but actually had Proteus Syndrome (Tibble and Cohen 1986). The paper was subtitled: "The Elephant Man Diagnosed". Proteus Syndrome has become understood to be the accurate diagnosis. Despite this, confusion and inaccuracies persist about neurofibramotosis and Proteus syndrome (Spring 2001; Proteus Syndrome Foundation 2018). Legendre et al. (2011) document numerous examples in the medical literature where Merrick is still wrongly associated with neurofibromatosis; this causes unnecessary suffering for those diagnosed with neurofibromatosis. Additionally, in this century, a new flurry of activity regarding diagnosing Merrick has surfaced. In 2001, a British researcher proposes Merrick had, in fact, suffered from both Proteus Syndrome and neurofibromatosis; while a Discovery Channel documentary in 2003 attempted to use DNA sampling and genealogical research to definitively diagnose him, and even reconstruct his face without disfigurement (Highfield 2003).

In the reproductive world, a few things are simmering that make the Lynch film compelling to its 1980 audience. Mastering the science of manipulating gametes leads to the birth of the first test tube baby (Dow 2017), which subsequently gives birth to the fertility treatment subspecialty and industry. Prenatal diagnosis also advances, which some authors have argued, is a euphemism for "eugenics" as early diagnosis of fetal anomalies or chromosomal accidents such as Down's Syndrome can give women the option of terminating their pregnancies. Women who were having pregnancies at older ages were recommended for amniocentesis. This was combined with a new technique in 1976 of DNA analysis to look for genetic disorders in developing fetuses. In 1977, "real-time" ultrasound equipment became available, which enabled healthcare providers to view a live image, leading to more prenatal diagnosis. Other techniques such as sampling cord blood, or using maternal alphafetoprotein to diagnose fetal conditions were not employed until the 1980s, but the research to perfect prenatal diagnosis—a societal desire—was well underway. This was in response to changing attitudes about pregnancy, and a social shift to delay childbearing. Women by 1980 had begun to shift their priorities and focus on their careers; this meant that, while, "biology was not destiny", there were still many increased risks of birth defects with delaying pregnancy and childbirth. Down's syndrome increased precipitously in older pregnancies, for example. The introduction of reproductive endocrinology and fertility treatment as a subspecialty begins to take shape around this timeframe. The story of Merrick, who is a catastrophic example of a "freak of nature," strongly resonates with female audiences. Lynch's use of Merrick's mother as a silent character in the film, speaks to women of this timeframe and beyond.

Clinical Ethics Themes in *The Elephant Man*

The Elephant Man can be theoretically approached as a film about autonomy and "Respect for Persons" (see previous sections) or "beneficence"—which is my approach here. I argue that this is really about beneficence—maximizing benefit while minimizing harm—that indeed supports the principle of autonomy. A beneficent plan of care must be guided by a patient's preferences, if known. If not known, then all we have to guide us are our own sense of what is in a patient's best interests, using a "reasonable person" standard. What would a reasonable person want in a particular clinical scenario? Beneficence thus replaces autonomy when we determine patients do not have decision-making capacity (see previous section) or are not competent to make decisions. Decision-making capacity also includes patients' ability to communicate their wishes, which is a definite challenge for Merrick on film, and in real life. Merrick is initially determined to be an "imbecile" in Lynch's treatment of *The Elephant Man*.

Lynch's storyline focuses on Treves as the central character, with the patient being in a more secondary role. At first, we see that Treves is interested in displaying Merrick as a medical curiosity, which will bring him a publication and some career perks. Later, when Merrick is brought to Treves' for medical care, his moral struggle with how to beneficently care for his patient truly begins. The film is really about the *doctor's dilemma* rather than Merrick's right to self-determination. The dilemma appears with the "capacity assessment" scene, in which Treves is hoping to prepare the patient to meet with the head of the hospital, Carr Gomm (played by John Gielgud). By prepping Merrick with some words to say, Treves is hoping for Gomm to see that Merrick is a human being who can speak, deserving of the hospital's mercy, and therefore, hospice and palliative care. In a pre-National

Health Service England, hospitals were for-profit institutions, and were not in the business of charity. Treves' plan backfires. Gomm recognizes that Merrick is merely parroting what he was taught to say, but that that there is no real intellect there. The absence of intellect would make it easier for the hospital to discharge Merrick to the streets, which is where "imbeciles" and freaks in Victorian England resided, and not hospitals. Gomm is not without compassion; he says that he "prays to God" Merrick is an imbecile. When Merrick clearly overhears Gomm's and Treves' discussion about his fate, he finally reveals his intellect and recites a prayer, revealing his deep intellect. It is at this moment, that the true ethical dilemma unveils. Treves and Gomm recognize that Merrick's intellect, trapped in his deformed body, represents unparalleled human suffering that deserves hospice, sympathy, and empathy. This recognition, in a Victorian timeframe, is quite admirable. History records that Carr Gomm contacted other institutions and hospitals more suited to caring for Merrick, but none would accept him. Carr Gomm then sought extramural support for Merrick. He wrote a letter to the daily newspaper outlining Merrick's case and asked for donations to the hospital to care for Merrick. Gomm wrote: "Some 76,000 patients a year pass through the doors of our hospital, but I have never before been authorized to invite public attention to any particular case, so it may well be believed that this case is exceptional (Gomm 1886). In Lynch's treatment, we see this resource allocation tale in some detail as well, in which he also shows the Princess of Wales personally paying a visit to Gomm and asking him to provide the necessary hospice. Lynch's film also provides an opportunity to discuss consent, as Treves indeed checks in with Merrick and asks him if permanent hospice is his desire.

In a present day setting, the ethical dilemma of the costs of palliative care, or long-term care for the disabled, is a natural discussion that is generated by the film. Carr Gomm has a different dilemma as a hospital administrator than Treves does as Merrick's attending physician. *How are we going to pay for this?* These questions have not gone away.

Treves moral dilemma changes midway into the film, however. As Merrick becomes a permanent resident at the London Hospital, and receives visitors from the upper class, he starts to see himself as an agent of exploitation, and possibly harm. Here is where maleficence and beneficence meet. Treves begins to suffer from moral distress over what he has done; the moral distress emanates from an epiphany in a scene in which he is found alone crying by his wife. Here, he reveals to his wife his moral distress about whether his initial intentions were truly "beneficent" or merely self-serving. This is a core discussion to have when showing the film; particularly in the United States healthcare system that is unabashedly self-serving as the debate over universal healthcare rages. In the end, it is difficult to say whether Treves was a "good man or a bad man". The harsh reality is that the practice of medicine in academic medical centers both exploits and serves patients. Medical students are taught to see patients as a means to their training every day. The "interesting" cases are dissected in case conferences and grand rounds. We enroll patients in clinical trials and research to serve both science and the researcher's career so s/he can achieve tenure, funding, and notoriety. Treves' soliloquy, in which he compares himself to the Freak Show manager, is just as relevant in a twenty-first century academic medical center as it is in a Victorian London hospital. We must note, however, that notwithstanding concerns of exploitation of this patient, Treves was a caring and humanistic physician who visited with Merrick often, and spent many hours a week with him as Waldron (1980) reminds us.

Conclusions

There are wider ethics discussions generated by this film, which include disability ethics; eugenics and prenatal diagnosis; research ethics and biobanking. For example, Treves took plaster casts of Merrick's head and limbs upon Merrick's death, as well as numerous skin samples. Many of these samples still remain in a pathology archive at the Royal London Hospital. Merrick's remains are endlessly examined. Do we have a right to do this? How do we feel about posthumous research on human subjects that may not have provided consent? Ultimately, however, within a *clinical ethics* context, this is a film about beneficence and asking what constitutes a benefit and what constitutes harm. Treves, in Lynch's treatment of the film, isn't sure. In caring for persons in need of hospice, long-term or palliative care, neither are most physicians.

Film Stats and Trivia

- Nominated for the following Academy Awards: Best Picture, Actor (John Hurt), Art Direction, Costume Design, Director, Film Editing, Original Score and Adapted Screenplay.
- Won the British Academy of Film and Television Arts (BAFTA) award for Best Film, Best Actor (John Hurt) and Best Production Design, and was nominated for: Direction, Screenplay, Cinematography, and Editing.
- John Hurt's makeup was made from casts of Merrick's body. The final make-up was devised by Christopher Tucker, which inspired the creation of a new Academy Award category for Best Make-up.
- In addition to writing and directing the film, Lynch provided the musical direction and sound design. During its depiction of the final moments of Merrick's life, the film uses "Adiagio for Strings" by Samuel Barber.
- Actor Frederick Treves, great-nephew of the surgeon, appears in the opening sequences as an Alderman trying to close down the freak show.
- The film was a critical and box office success, grossing $26,010,864 (USA).

From Theatrical Poster

Directed by David Lynch
Produced by Jonathan Sanger, Stuart Cornfeld and Mel Brooks
Screenplay by Christopher De Vore, Eric Bergren, David Lynch
Story by Sir Frederick Treves, Ashley Montagu
Starring Anthony Hopkins, John Hurt, Anne Bancroft, John Gielgud, Wendy Hiller
Music by John Morris
Cinematography Freddie Francis
Editing by Anne V. Coates
Studio Brooksfilms
Distributed by Paramount Pictures (USA) EMI (International) Release date(s) October 3, 1980 (1980-10-03)

References

Ablon, Joan. 1995. The Elephant Man as self and other: The psychosocial costs of a misdiagnosis. *Social Science and Medicine* 40: 1481–1489.

Aguirre, Robert. 2010. Review of the spectacle of deformity: Freak shows and modern british culture. *The American Historical Review* 115: 1530–1531.

Alexander, Caroline. 2007. Faces of war. *Smithsonian* 37: 72–80.

A remarkable case of double monstrosity in an adult. 1866. The London Lancet. 1:71–73.

Biesecker, Leslie G., R. Happle, John B. Mulliken, Rosanna Weksberg, John M. Graham Jr., Denis L. Viljoien, and Michael Cohen Jr. 1999. Proteus syndrome: Diagnostic criteria, differential diagnosis and patient evaluation. *American Journal of Medical Genetics* 84: 389–395.

Canby, Victor. 1980. The Elephant Man (Review). *New York Times*, October 3.

Carswell, Heather. 1982. Elephant Man had more than neurofibromatosis. *Journal of the American Medical Association* 248: 1032–1033.

Cohen Jr., Michael M. 1986. The Elephant Man did not have neurofibromatosis. *Proceedings of the Greenwood Genetic Center* 6: 187–192.

Cohen, Michael M. Jr. 1988a. Further diagnostic thoughts about the Elephant Man. *American Journal of Medical Genetics* 29: 777–782.

Cohen, Michael M. Jr. 1988b. Understanding Proteus syndrome, unmasking the Elephant Man, and stemming elephant fever. *Neurofibromatosis* 1: 260–280.

Cohen, Michael.M. Jr., and Patricia.W. Hayden. 1979. Penetrance and variability in malformation syndromes. *Birth Defects* 15: 291–296.

Darke, Paul.A. 1994. The Elephant Man: An analysis from a disabled perspective. *Disability & Society* 9: 327–342.

Dow, K. 2017. 'The men who made the breakthrough': How the British press represented Patrick Steptoe and Robert Edwards in 1978. *Reproductive Biomedicine and Society Online* 4: 59–67. https://www.sciencedirect.com/science/article/pii/S2405661817300199.

Durbach, Nadja. 2009. *The Spectacle of Deformity: Freak Shows and Modern British Culture*. Berkeley: University of California Press.

Elof, C. 2018. Eugenics Archive. Dolan DNA Learning Center, Cold Springs Harbor Labaratory. http://www.eugenicsarchive.org/html/eugenics/essay2text.html. Accessed 25 Feb 2018.

Fiedler, Leslie. 1996. Preface. In *Freakery: Cultural Spectacles of the Extraordinary Body*, ed. Rosemarie Garland Thomson, xiii–xiv. New York: New York University Press.

Friedman, Lester D. 2000. Medicine and the arts. *Academic Medicine* 75: 448–449.

Garland Thomson, Rosemarie (ed.). 1996. *Freakery: Cultural Spectacles of the Extraordinary Body*. New York: New York University Press.

Gerber, David A. 2003. Disabled veterans, the state, and the experience of disability in western societies, 1914–1950. *Journal of Social History* 36: 899–915.

Gomm, Francis C. 1886. The Elephant Man. *The Times*, December 4th (Note: The original letter from the newspaper can be seen at the following link. https://blackcablondon.net/2012/03/13/joseph-merrick-the-elephant-man-part-two/. The letter text is also preserved here: http://www.lettersofnote.com/2016/08/the-elephant-man.html).

Highfield, Roger. 2003. Science uncovers handsome side o elephant man. *The Telegraph*, July 22. http://www.telegraph.co.uk/news/uknews/1436744/Science-uncovers-handsome-side-of-the-Elephant-Man.html.

Howell, Michael, and Peter Ford. 1980. *The True Story of the Elephant Man*. London: Allison and Busby.

Hubbard, Ruth. 1997. Abortion and Disability: Who should and should not inhabit the world In *The Disability Studies Reader*, ed. Lennard J. Davis, 74-86. London: Routledge.

Huddelston, Tom. 2008. Interview with David Lynch. Time Out. http://www.timeout.com/film/features/show-feature/5443/david-lynch-interview.html. Accessed 24 Feb 2018.

Johnson, Jeff. 2003. Pervert in the pulpit: The puritanical impulse in the films of David Lynch. *Journal of Film and Video* 55: 3–14.

Kochanek, Lisa A. 1997. Reframing the freak: From sideshow to science. *Victorian Periodicals Review* 30: 227–243.

Legendre, C.M., C. Charpentier-Cote, R. Drouin, and C. Bouffard. 2011. Neurofibromatosis Type 1 and the 'Elephant Man's" disease: The confusion persists: An Ethnographic Study. *PLoS ONE* https://www.ncbi.nlm.nih.gov/pmc/articles/PMC3036577/.

MacKenzie, Donald. 1975. Eugenics in Britain. *Social Studies of Science* 6: 499–532.

Medical News. 1890. Death of the Elephant Man. *British Medical Journal* 1: 916–917.

Meet Louise, the world's first test tube baby. 1978. *Evening News,* July 27.

Montagu, Ashley. 1971. *The Elephant Man: A Study in Human Dignity*. New York: Outerbridge and Dienstfrey.

Paramount Pictures. 2001. *The Elephant Man Revealed*. December 11.

Proteus Syndrome Foundation. 2018. What is Proteus Syndrome? https://www.proteus-syndrome.org/newly-diagnosed.html. Accessed 25 Feb 2018.

Randell, Karen. 2003. Masking the horror of trauma: The hysterical body of Lon Chaney. *Screen* 44: 216–221.

Resta, Robert G. 1997. The first prenatal diagnosis of a fetal abnormality. *Journal of Genetic Counseling* 6: 8184.

Roe v. Wade, 410 U.S. 113 (1973).

Sparks, Christine. 1980. *The Elephant Man*. New York: Ballantine Books.

Spring, Paul. 2001. The improbable Elephant Man. *Biologist* 48: 104.

The Mutter Museum. 2018. History of the Mutter Museum. http://muttermuseum.org/about/history/. Accessed 25 Feb 2018.

Tibbles John, A.R., and Michael M. Cohen. 1986. The Proteus syndrome: The Elephant Man diagnosed. *British Medical Journal* 293: 683–685.

Toulmin, Vanessa. 2009. Review of Victorian Freaks: The social context of Freakery in Britain. *Victorian Studies* 51: 740–741.

Treves, Frederick. 1884. Congenital deformity. *British Medical Journal* 2: 1140.

Treves, Frederick. 1885a. A case of congenital deformity. *British Medical Journal* 1: 595.

Treves, Frederick. 1885b. A case of congenital deformity. *Proceedings of the Pathalogical Society of London* 36: 494–498.

Treves, Frederick. 1923. *The Elephant Man and Other Reminiscences*. London: Cassell and Co.

Waldron, Gillian. 1980. Sir Frederick Treves and the Elephant Man. (Letter to the Editor). *The Lancet* 316: 42.

Weber, F. Parkes. 1909. Cutaneous Pigmentation as an incomplete form of Recklinghausen disease. *British Journal of Dermatology* 21: 49–53.

Weber, F. Parkes. 1930. Periosteal neurofibromatosis, with a short consideration of the whole subject of neurofibromatosis. *Quarterly Journal of Medicine* 23: 151–165.

Wilkie, Theodore F., and J. Milton Rabson. 1979. The Elephant Man—A tragic syndrome. *Aesthetic Plastic Surgery* 3: 327–337.

Wilson, Karina. 2001. On Tod Browning's Freaks. Horror Film History. http://www.horrorfilmhistory.com/index.php?pageID=freaks. Accessed 24 Feb 2018.

Pediatric Ethics and the Limits of Parental Authority: *Lorenzo's Oil* (1992)

9

Like every film in this section, *Lorenzo's Oil* is based on the actual case of Lorenzo Odone (1978–2008), who at the age of 6, was diagnosed with a rare genetic disease, known as adrenoleukodystrophy (ALD), an x-linked recessive disease passed from the carrier mother to her child. This film is about the most extreme form of ALD in boys known as "childhood cerebral" or "childhood cerebral dymyelinating ALD", affecting only males from 3–10 years of age. Here, the male's cerebral white matter is slowly stripped of its myelin (demyelination), resulting in a severe neurode-generative decline that leads to a vegetative state usually within a couple of years after diagnosis. Demyelination occurs in this case because of an abnormal accu-mulation of saturated very-long-chain fatty acids (VLCFA) in the serum and tissues of the central nervous system, which sets off an abnormal immune response that leads to demyelination. When Lorenzo Odone was diagnosed in 1984, the disease had just recently been described in the medical literature and there was no treatment at all; his parents, Augusto and Michaela Odone, would not accept this, and developed an innovative therapy known as "Lorenzo's Oil" which is a mixture of oleic acid and erucic acid. The efficacy of Lorenzo's Oil is discussed further under History of Medicine, and is used in one type of ALD under certain circumstances. Lorenzo, himself, seemed to mildly improve on it, and he lived until age 30, which represents the longest lifespan of a boy with ALD. But, the burning question is: at what cost?

This film deals with pediatric ethics themes, and whether the parents are really making decisions that are "beneficent" for their child, using a "best interests of the child" framework. This film looks at the issue of weighing therapeutic benefits against potential harms, which include psychosocial harms. The case of Lorenzo's Oil is a true test case for beneficence, as some would argue that the parents' decisions were in their child's best interests; some would argue they caused greater suffering and perhaps should have been removed as decision-makers. This film also hones in on "innovative therapy" and a patient's right to "try" something new and outside the box when there is no available treatment. Finally, this film asks us to consider when

M. S. Rosenthal, *Clinical Ethics on Film*,
https://doi.org/10.1007/978-3-319-90374-3_9

is an innovative therapy "research" that requires a proper research protocol? This film had become popular, too, as a "teaching science" film (which I will not focus on here), as the parents teach themselves biochemistry in order to think about how they can find a treatment. The Odone's work on Lorenzo's Oil, and later, on the "Myelin Project" (see under History of Medicine) is laudable and inspiring, but it is also a *cautionary* tale about the limits of parental authority, as there are uncomfortable truths about how much their work actually helped their own child. As in many medical research stories, their discovery has certainly helped other children, although there is still controversy over the efficacy. It is not clear what level of awareness or assent Lorenzo had in being the recipient of his parents' experimental therapy for him, and at what point they crossed a line. The Odone's own lives were consumed by their work for Lorenzo, and indeed many parents would not, and perhaps should not, be expected to respond to a child's devastating illness in the same way. When teaching this film as a pediatrics ethics case study, some of the scenes depicted, if true, would have warranted a call to Child Protective Services. Thus, teaching *Lorenzo's Oil* as a clinical ethics film requires a distancing from the romanticism of the parents' heroic efforts to save their child, and a harder look at whether any of this was in the child's best interests. When this film was first released in 1992, I recall seeing it onscreen and coming away with the sense that the parents were heroes. When revisiting the film as a clinical ethicist, I remain uncomfortable, and so should the classroom. This is what makes this film a true case for a beneficence dilemma. It's really not clear whether we harmed, or helped, Lorenzo.

The Making of *Lorenzo's Oil:* International Intrigue

The story of Lorenzo Odone's plight was originally captured on Italian television in a 1990 made-for-television Italian drama entitled: *Voglia di vivere* ("Desire For Life") directed by Lodovico Gasparini (Fumarola 1990). In this original screenplay, which changed the names of all the characters, the story was based on the actual life events of the Odones. Augusto Odone (1933–2013), Lorenzo's father, was a native Italian born in Rome, and was very active in the script's development. However, this was a distinct teleplay from the American film (1992), which had nothing to do with this nascent version of the story. The telefilm premiered on Italy's Channel 5 for its "Film Dossier" series. An Italian article written on the night of its premiere states: "Lorenzo's Oil in Italy is used by the Bambin Gesù Hospital in Rome and at the Pavilion Carlo Besta of the Polyclinic in Milan." Augusto Odone, who is highlighted in the article, stated the following (translated from Italian): "Together [my wife and I] speak five languages ...and our classical studies have allowed us to address the neurobiology texts.... Science has its rhythms and its methods...I had to fight against my son's death and I was not going to resign myself" (Fumarola 1990).

After the Italian telefilm premiered, the channel next aired a short documentary following the film, which featured Augusto, and Drs. Marco Cappa and Enrico Bertini, an endocrinologist and neurologist respectively, at the Bambin Gesu

Hospital in Rome, who both specialized in the treatment of adrenoleukodystrophy (Fumarola 1990).

The Italian telefilm notwithstanding, two other filmmakers became interested in the Odone's story around this time. One was Italian director, Franco Amurri, who had been in touch with the Odones as early as 1988. Amurri began researching for a potential film using the couple's audiotaped diary about their experiences. Amurri, who was dating Susan Sarandon at the time (and had a daughter with Sarandon), gave her the Odone tapes to listen to, as he thought about the story and its potential for a feature film (Wuntch 1993).

Unbeknownst to Amurri, George Miller, an Australian physician-turned film-maker (who had directed *The Witches of Eastwick* costarring Sarandon), independently read about the Odone's case in *The London Times* around 1990 (Dutka 1992), and contacted the family himself because he wanted to make an American feature film about their plight. Miller's film was created from an original screenplay that was not based on the earlier Italian version, nor was Miller privy to Franco Amurri's interest in the story (Wuntch 1993).

As the American film project was in process, the Odone's story began to be more widely written about in the American media. For example, a major story in *People* magazine about the Odones was published in 1991 (Ryan 1991), discussed further. Alongside the Odone's struggle to find a cure for a rare disease for their son, many families were dealing with another disease: AIDS. The story of the Odone's battle to find a cure within a medical bureaucracy had universal appeal in light of patients with AIDS finding themselves in a similar race for time.

The *Los Angeles Times* reported the following about the Odones (Dutka 1992):

> An Italian economist working for the World Bank and his Irish- American linguist wife had together discovered a means of reducing the fatty acids that caused adrenoleukodystrophy (ALD), the nervous system malfunction that constituted a certain death sentence for their son. Today, eight years after the diagnosis, 14-year-old Lorenzo is severely disabled, awaiting further breakthroughs. And the "oil" that ultimately saved his life? Like a host of drugs that show promise in combatting AIDS, it is unavailable for mass distribution pending FDA approval.

Miller had initially cast Michelle Pfeiffer for the role of the mother, Michaela Odone, but Pfeiffer was unsure if she could pull it off. Enroute to Los Angeles to meet with Pfeiffer, Miller ran into Susan Sarandon, who was seated near him on the flight. It was on this flight that Miller learned that Sarandon was already very familiar with the case and Odone family because of her relationship with Amurri. Webster and Sarandon spent the entire flight to Los Angeles discussing the case, and their shared passion about it; it was clear to both of them that Sarandon should play Michaela. Sarandon then met with Michaela in person on several occasions to understand her better, and play her; Sarandon also became the celebrity spokes-woman for the Myelin Project, founded by the Odones. Nick Nolte, who was cast as Augusto, initially pulled out of the film due to personal problems, and was briefly replaced by Andy Garcia. But Garcia was too young for the part, and was, himself, not comfortable in the role. Nolte returned to the part after all, and although criticized for his labored Italian accent in the film, actually embodied Augusto in many other ways (Wuntch 1993).

As it turned out, the most problematic actor was one of the six children who played Lorenzo (various ages were shown): Zack O'Malley Greenburg, whose parents wreaked havoc on the set. According to Miller: "They were shameless… They wanted to sell a screenplay to us. They wanted credit as coaches. They needed us for self-esteem because they were going through a divorce. They used the film shamelessly" (Wuntch 1993). Ultimately, the production team worked through the problems with Greenburg's parents. It's also important to note that the Odones did not have script approval.

Synopsis

This is the dramatized case study of Lorenzo Odone, who suffered from the genetic disease (x-linked), adrenoleukodystrophy—ALD. Lorenzo was diagnosed with this disease in 1984 when he was 6. His parents were highly trained professionals; his father, Augusto Odone, was an economist for the World Bank, and his mother, Michaela, was a trained linguist. When they discover their son has this devastating, terminal disease, they decide to research a treatment on their own, and invent "Lorenzo's Oil". There is a lot of biochemistry about the oil and the disease thoroughly explained in the film. Not everyone supports or agrees with the parents; founders of an organization for children with ALD find the parents reckless, while the worlds' expert in the disease is dubious about whether the parents are doing more harm than good. The viewer must decide in the end whether the parents' sacrifice, their discovery, and Lorenzo's quality of life, add up to a beneficent result.

The Odones: Their Story and Social Location

Augusto and Michaela Odone, a highly educated couple, were living a fairly privileged life in the late 1970s when their child was born. Michaela Murphy Odone (1939–2000) was a linguist, and had completed a degree in French in 1962 in Washington, D.C. and then spent several years studying in France, some of it as a Fulbright scholar at the University of Grenoble. She worked as a linguist in both Paris and New York prior to meeting Augusto. Augusto Odone (1933–2013) worked as an economist for the World Bank, and had been previously married to a Swedish woman with whom he had two grown children. He divorced his first wife before meeting Michaela. The couple had traveled all over the world, and had Lorenzo when Michaela was 39—quite late, considering the timeframe, which was prior to fertility treatments or IVF. However, Michaela's age had nothing to do with her passing on a genetic disease to Lorenzo (see under History of Medicine). She was a carrier for a disease that had only very recently been identified (Saxon 2000; Vitello 2013).

The Odones were an international couple that led an unusual and fascinating life. Augusto came of age in post-war Italy, and spoke several languages. According to his son in-law, who informed one of his obituaries (The Economist 2013):

> Augusto came from a family of stubborn, quixotic, maddening fighters. His mother—a pioneer of Italian domestic science—had browbeaten Mussolini into allowing her books to be published. His father, Angelo, an army general with a British medal, was a leader of the anti-fascist resistance in wartime Rome; young Augusto narrowly escaped execution for hurling stones at passing German tanks.

He, too, was a Fulbright scholar who was an expert in developing economies for the World Bank. In 1978, the couple was based in Washington, D.C., where they had Lorenzo, but due to work demands, Augusto moved the family to the Comoros, a French-speaking island nation in the Indian Ocean. Augusto's work involved developing the Comoros economy, and Michaela busied herself there with running an informal clinic, and distributing medicines donated by charitable U.S. organizations. She helped Augusto write the Economic Plan of the Comoros, which was a United Nations-financed and World Bank-executed project (Saxon 2000; Vitello 2013). Thus, the social context for the Odones was global development work and activism; they were "citizens of the world" more than they were citizens of any single country. And although they were essentially the same demographic as the Beat Generation, which has relevance to other films in this book (see Chaps. 4 and 6), the Odones were more influenced by global post-World War II events more than specific cultural events in the United States. Their worldview of disease was global and not local. They were very individual and non-conformist; they lived "outside the box" and thus, when confronted with Lorenzo's illness in 1984, their response was "outside the box" as parents of a sick child. Lorenzo began to show signs of ALD (emotional outbursts, speech problems, etc.) at age 5, when he was in pre-school, but it took some time for a definitive diagnosis of ALD. When the Odones learned that ALD was an "orphan disease" with no real treatment, a life expectancy of less than two years, and no serious allocation of research funds, they decided to treat the disease as though it were a foreign country and language, and immersed themselves in the language of the disease: they learned biochemistry, and acted "globally" by financing an international conference/roundtable of any relevant experts in the disease so they could gain more knowledge. They also had difficulty with what they viewed as parochial or more cloistered attitudes about parental decision-making. They found that the organizations for parents of ALD patients were primarily in the service of the parents' psychosocial needs rather than the children, which they viewed as selfish. However, pediatric ethicists may argue that support for a dying child's caregivers is actually in service to the child. As such, the Odones badly clashed with traditional American medical culture and views, including guidelines for clinical trials and research (see under Clinical Ethics Issues).

Eventually, Augusto forged a partnership with Dr. Hugo Moser, co-published medical journal papers with him, and was awarded an honorary doctorate by the University of Stirling in Scotland (Childs 2013) for his contributions on "Lorenzo' Oil". Moser was the world's expert in ALD, and called "Dr. Nicholais" in the film version, which Moser later argued was not an accurate portrayal of his involvement (see under History of Medicine).

Within the American context, the Odones were dealing with Lorenzo's illness during roughly the same timeframe as the AIDS epidemic, coinciding with dramatic funding cuts levied during the Reagan Administration. In 1985, the Reagan Administration reduced the number of biomedical research grants by 23% (DeCoursey 1985), which had taken a toll on the research community. With far more desperate patients requiring AIDS research funding, ALD was an unfunded orphan disease, which the Odones took up as their personal cause. One can argue that their involvement in research was a tremendous good, but they entangled it with decision-making around their own child, which is where the beneficence dilemma resides (see under Clinical Ethics Issues). The research efforts of the Odones need to be separated from their parental decision-making for Lorenzo. In fact, a similar parents-turned-research story surrounding progeria was chronicled in the HBO documentary, *Life According to Sam* (2013) in which the parents do just that. They are driven to research a cure based on their son's plight with progeria, but are very careful to separate decisions as research investigators from decisions for their son, based on current standards of care. Some of this, can be argued, was perhaps a "lessons learned" from the Odone story.

The United Leukodystrophy Foundation

Although their names are changed in the film, another couple featured in the Odone's story is that of Ron and Paula Brazeal. The Brazeals, based in Sycamore, Illinois, founded the United Leukodystrophy Foundation in 1982. They had lost two boys to ALD, one who died 18 months after diagnosis, in 1978, and one who died six years after his diagnosis, in 1988, after six years of being bedridden and suffering similarly as Lorenzo. This couple was open to any experimental therapies that were recommended by the medical community for their second son, but they were skeptical and resistant to the Odone's "oil" because it had not been FDA approved or recommended by the medical community as efficacious. The couple's skepticism represented rational decision-making based on classic beneficence frameworks: maximizing benefits and minimizing harms. The Odones and Brazeals badly clashed because the Brazeals would not give their Foundation's endorsement to the oil, which was regarded as an innovative therapy, before it went through more testing. This was entirely reasonable within the American medical construct. But within a global health construct, in which "anything" may be better than "nothing", it seemed unreasonable to the Odones.

The Brazeals were also dubious about prolonging their child's suffering, and stated: "We felt at the time that it was not a cure and would not improve him... The most we could hope for was that it would prolong him in the (severely degenerated) state he was in" (Storch 1993). At the time the Brazeals made this statement, Lorenzo was 14, still paralyzed and in need of constant care, but could communicate with his parents primarily by blinking or wagging a finger; some would call that quality of life while others would call it prolonging suffering. It remains unclear whether the oil given to Lorenzo helped to stop progression of his disease (it could not reverse the damage already done), which is discussed under the History of Medicine section. As the founders and leaders of a legitimate non-profit disease foundation, which is accountable to the public trust, the Brazeals indeed had a responsibility to ensure that their foundation endorsed research that followed proper protocols for clinical research, which had just become recently codified in *The Belmont Report* (see under History of Medicine). At the same time, there was a responsibility to ensure that research on children went through more rigorous review, as they were a vulnerable population. The film's depiction of the Brazeals tends to bias the viewer toward the Odone's worldview, as the film is from the Odone's perspective. However, the Brazeals arguments with the Odones are authentically presented and the film still preserves their point of view and dilemma. When teaching, it's important to ensure that the Brazeals story (or the story of their fictionalized characters) is unpacked. The Brazeals actually did praise the film for its accurate portrayal of ALD, but felt their perspective was minimized because they championed palliative and hospice protocols for their suffering children (Storch 1993). In the end, most pediatric ethicists would support the Brazeal's decision-making as far more rational than the Odone's, which some could argue was complicated by the couple's pathological grieving. Regarding the film's depiction of the Brazeals and their foundation (disguised as the "Adrenoleukodystrophy Foundation"), Michaela Odone stated (Storch 1993):

> Frankly, we find it is very symbolic of support groups, particularly pediatric ones, that feel their mission in life is to support and uphold the medical Establishment... I would like the movie to make patients and parents a little bolder in questioning dogmatic statements from doctors. For [potential treatments] to get sidetracked in this ridiculous controversy is to do a terrible disservice. I'm sorry if people are offended by their characterizations [in the film], but there are liberties taken with my character and my husband's, too.

The Odones remained active members of the United Leukodystrophy Foundation, and the Odones acknowledged that the foundation helped many parents with their distress.

It's important to note that foundations such as the United Leukodystophy Foundation represented patient support groups pre-internet, and played critical roles in patient education, and patient/family-led psychosocial support in this timeframe. These groups were often the only sources of patient-accessible literature and materials for rare or orphan diseases. Such foundations served as important conduits for research funding, community-led activism and conferences. Typically, these non-government organizations had special tax status, a medical advisory board, board of directors, and were required to act in accordance with strict regulations regarding research and support activities. It's important to ensure that in teaching, students understand the Odones were, in fact, outliers, and the Brazeals had greater responsibilities to their members, board and the public trust. The Brazeals were tasked with ensuring their community of parents and children were protected. In a post-internet environment, virtual patient groups sprung up all over the place, and often had no defined accountability. Patient listservs often became unruly sources of misinformation based on unverified or untested claims. In some ways, the Brazeals represent a period in patient-support history where non-profit activities were in many ways more responsible (Storch 1993).

The Odones' Psychosocial Experience and Decision-Making Consequences

The Odone's psychosocial lives essentially revolved solely around Lorenzo to the point where they had no social functioning outside of Lorenzo or in service of researching his disease. This is accurately depicted in the film, when Michaela's own sister tells her that "there has to be a life beyond Lorenzo" and Michaela throws her out of the house for her suggestion that she should consider her own needs. The real, and portrayed, Michaela spent every waking hour at Lorenzo's bedside for the rest of her own life. Michaela died from lung cancer at age 61, and Lorenzo outlived her, which was likely her greatest personal achievement: outliving her child. Michaela was a "mother tiger" archetype who was fiercely devoted to, and protective of, her child. She was unable to allow herself any personal pleasure once her son was diagnosed. This is not what reasonable mental health professionals would advocate for any parent/caregiver of a sick child, however. As for Augusto, he took an early retirement in 1987 from the World Bank to spend all of his time and energy on Lorenzo and research into ALD. Unlike Michaela, Augusto outlived Lorenzo; Lorenzo died the day after his 30th birthday in 2008. Augusto lived five more years until 2013, but had spent all his time at Lorenzo's bedside (with the exception of Lorenzo-related activities) until Lorenzo died from choking on his saliva—a continuous risk of his condition.

  Lorenzo's quality of life was questionable, too. He lived mostly motionless and blind from around age six to his death, but when he began to take the oil his parents invented, he apparently regained the use of his fingers, and communicated by wiggling fingers and blinking. He remained blind and motionless, unable to feed himself or essentially have much quality of life. He did have awareness, and appears to have had appreciation of being read to, and spoken to. It is inconclusive whether his starting the oil helped to ease or stop progression of some of his symptoms (see under History of Medicine). Assuming the oil played a role, did his parents prolong his suffering? Many clinical and pediatric ethicists would argue that Lorenzo might have been better off with a hospice/palliative care protocol that allowed natural death without interference. In the absence of the oil, his parents still insisted on all aggressive measures to keep him alive, such as feeding tubes, and continuous saliva suctioning round-the-clock. Other parents made different decisions in the same disease context, including the Brazeals, who allowed natural death for their ALD-stricken children, given that most reasonable people would not accept Lorenzo's life as an acceptable quality of life for the child or family. In an extraordinary, and highly questionable move, the Odones actually brought a young man from Africa (who Lorenzo had befriended in the Comoros) to come to the U.S.

to be a "nursing aid" to Lorenzo. In an interview after Michaela's death Augusto "conceded…that he had sometimes wondered if [Lorenzo had] enough of a life to justify the extraordinary lengths to which he and his wife had gone" (Vitello 2013). According to Augusto's daughter, Christina: "Lorenzo never regained his faculties… If you had ever walked into the room and seen how Lorenzo responded to the way my father and Michaela embraced him in life, wrapped him in love, you would see he was a living being who knew he was loved. That's what they gave him, but it was very difficult" (Vitello 2013).

There is a romanticization of the parental decisions in this case, which are evident from this 1991 *People* magazine spread on the couple (Ryan 1991):

> Now Lorenzo is 13 and still remarkable. He loves opera and Dickens and Kipling and still understands three languages. But he has not spoken in six years. He has a handful of attendants and nurses for round-the-clock care. His friend Oumouri Hassane, now 26, came from the Comoros in 1985 to help. They watch him constantly, careful to notice when he wiggles his fingers or blinks or moves his head. These are the ways in which he communicates.
>
> … Had events taken their normal course, Lorenzo would have died in 1986 or 1987—two years after he was stricken with a rare genetic disease called ALD (adrenoleukodystrophy), which strikes young boys and slowly paralyzes, then kills them. He is alive today because his mother and father, with virtually no scientific training, threw themselves into medical literature in a desperate effort to find a cure for his disease. They didn't find one, but against all probability, they did find a way to halt the progress of ALD and prevent susceptible children from developing the symptoms.

I would pose the argument to students when teaching this film that, perhaps, had Lorenzo died in 1986 or 1987, it would have been more beneficent, given the extraordinary toll and resources expended on a life that seemed more for the benefit of the parents than the child.

History of Medicine Context

There are three history of medicine contexts for this film: the history of ALD and its treatment, including the efficacy of Lorenzo's Oil; a review of the genomics involved in ALD, as it is genetically inherited; and a review of medical research involving children.

Types of ALD

There are three types of X-linked[1] adrenoleukodystrophy: (1) a childhood cerebral form, which is the focus of the film, *Lorenzo's Oil*; (2) an adrenomyeloneuropathy

---

[1]Note: To explain an x-linked genetic mutation, it's important students (especially non-science) understand the basics of why boys are more vulnerable: when women have an abnormal X chromosome, a second X chromosome could result in a normal phenotype. Boys and men do not

type, which is a milder form, and (3) another form resulting in Addison's disease only without any other neurological degeneration (U.S. National Library of Medicine 2018). The most severe form of the disease is a devastating diagnosis: the childhood cerebral form of ALD; this occurs in 45% of all ALD cases. "It is characterized by an inflammatory process that destroys the myelin, causing relentless progressive deterioration to a vegetative state or death, usually within five years" (The Stop ALD Foundation 2004). However, there are a variety of different phenotypes in males and females. The severe symptoms of the disease are mostly evident in males during childhood and adolescence. Females with the disease usually don't have any symptoms in childhood, but could have some milder neurological problems in adulthood, and rarely, dementia related to ALD. For the purposes of this discussion, I will focus on the childhood cerebral form of ALD, which only affects males, and which is passed from the carrier mother as a recessive gene. The film accurately depicts the presenting early symptoms of the disease and shows its progression: behavior disturbances; speech disturbances; problems with motor functions, until paralysis sets in. The risk of aspirating from the inability to swallow is the most common reason for death at that stage of illness.

When Lorenzo Odone was born in 1978, there were only a few cases of his form of ALD described in the literature. By 1981, Hugo Moser, the world's expert in the disease (portrayed as Dr. Nicholais in the film, and played by Peter Ustinov) published a groundbreaking paper on the mechanism of the pathogenesis of the disease, and its relationship to very longchain fatty acids (VLFAs). That same year, ALD was mapped to a segment of the X-chromosome, Xq28. When the Odones saga began, ALD was in its research infancy, and other than the ability to accurately diagnose the disease, and experimental protocols ranging from bone marrow transplants to a special diet (that was ineffective), there was absolutely no therapy other than palliative and hospice care until natural death. Lorenzo's oil, as a therapy, was formally introduced in 1989, with the publication of the definitive paper by Rizzo et al., co-authored by Augusto Odone (Rizzo et al. 1989). As the film depicts, in 1987, having taken early retirement, Augusto Odone investigated whether there could be a treatment in the form of two vegetable oils that could theoretically stop or delay the progression of ALD. He enrolled Lorenzo in a dietary therapy being investigated by Hugo Moser, which restricted saturated fats, and also added unsaturated oils, based on some preliminary research by William B. Rizzo, a researcher at the Medical College of Virginia in Richmond. Rizzo et al. published in 1986 that certain unsaturated oils could significantly lower VLCFA in ALD patients (Rizzo et al. 1986); this research formed the basis for pursuing Lorenzo's oil as a therapy. Moser based a clinical trial on this 1986 paper, which enrolled Lorenzo as one of the participants, and Augusto as one of the co-PIs, resulting in a 1987 paper (Moser et al. 1987).

To refine the oil concoction as a therapy, the Odones found an expert biochemist to help them: British scientist, Don Suddaby, from the Croda Universal Chemical

---

have a second X chromosome, and so if they inherit the x-linked disorder, they will have an abnormal phenotype (http://www.stopald.org/what-is-ald/).

Company. Suddaby was retired at that point, but took on the project (and plays himself in the film) to produce an edible mixture of olive and rapeseed oil extract, which became formally known as "Lorenzo's Oil". The Odones pursued the oil project based on guidance by Rizzo, (who was first author of a subsequent 1989 paper), who first suspected that certain oils might suppress build-up of VLCFA. Rizzo postulated that erucic acid and oleic acid could be the combination of oils that could be therapeutic in ALD. He stated to the *New York Times*: "The idea that monounsaturated fatty acids could be used therapeutically was derived entirely by me…It was not Mr. Odone's idea. To be honest, this whole thing started with that finding. Everything else was jumping off from that" (Kolata 1993). It's important to note that the Odones were instrumental in seeking answers from experts and raising the appropriate questions in order to spur thinking around a potential therapy. When they discovered there was a potential for erucic acid and oleic acid to be combined into therapy (based on Rizzo), they personally found companies to make it, and then tested the oil for safety in an $N = 1$ trial involving Michaela's sister, who was a carrier for the same gene; they sent her blood for testing and found that the oils indeed lowered the amount of fatty acids in her blood. Michaela stated in an interview: "We saw [Rizzo] had made an interesting observation about what happened in the laboratory but the consensus was that it would not work in vivo. The scientists said, 'We know it cannot work.' This was from Mount Olympus. Period." According to Rizzo: "The Odones have to deserve credit for not only finding companies that would synthesize both oils but also for pushing for oleic and erucic acid for their son" (Kolata 1993). The Odones decided to use it as an innovative therapy (see under Clinical Ethics Issues) on their own child, given there was no other treatment, even though there was a risk it could make things worse (Moser et al. 1987).

The promising 1989 abstract, which is the first definitive published paper on Lorenzo's Oil, states: "We conclude that dietary erucic acid therapy is effective in lowering plasma C26:0 to normal in ALD patients, and may prevent further demyelination in some mildly affected boys" (Rizzo et al. 1989). And thus began the controversy over the *efficacy* of Lorenzo's Oil, which remains as of this writing. It was never clear whether the oil helped Lorenzo, or whether Lorenzo's condition was among the clinically heterogenous presentations that included 12 other boys not on the oil, who also reached a clinical plateau. Moser stated to the *New York Times* in 1993: "I know of 12 patients who developed symptoms at the same time as Lorenzo who are still alive and in their teens or early 20s, even though they had no therapy; the fact that he survived is not an indication that the oil made a difference" (Kolata 1993).

Several papers by Hugo Moser and others followed the 1989 paper by Rizzo et al. (DiGregorio and Schroeder 1995; Poulos and Robertson 1994; Moser and Borel 1995; Lerner 2009), which made clear that there were questions about the efficacy of Lorenzo's Oil: research showed it was not effective in all forms of ALD (Aubourg et al. 1993). For example, it doesn't work at all in Adrenomyeloneuropathy (Aubourg et al. 1993), and was found ineffective in changing outcomes of symptomatic children with the cerebral form, despite the Odone's insistence in a

Letter to the Editor that it was premature to draw these conclusions at the time (Odone and Odone 1994; Auburg 1994; Rizzo 1994).

That said, the oil was not found to be harmful in any way, but there is harm associated with therapeutic misconceptions for desperate parents. However, there was finally consensus in 2007—seven years after Michaela's death and a year before Lorenzo's death – that the oil was effective in preventing symptoms in asymptomatic boys shown to have the genetic predisposition to the disease (Moser et al. 2007), bearing out the line in the film by Augusto, in which he realizes that perhaps all their toil was "for somebody else's kid":

> X-linked adrenoleukodystrophy (X-ALD) is a genetic disorder that damages the nervous system and is associated with the accumulation of saturated very long chain fatty acids (SVLCFA). Oral administration of "Lorenzo's oil" (LO), a 4:1 mixture of glyceryl trioleate and glyceryl trierucate, normalizes the SVLCFA levels in plasma, but its clinical efficacy and the clinical indications for its use have been controversial for more than 15 years. We review the biochemical effects of LO administration and the rationale for its use and present a current appraisal of its capacity to reduce the risk for the childhood cerebral phenotype when administered to asymptomatic boys and to slow progression of adrenomyeloneuropathy in patients without cerebral involvement. We also present current efforts to provide definitive evaluation of its clinical efficacy and discuss its possible role in the new therapeutic opportunities that will arise if newborn screening for X-ALD is validated and implemented (Moser et al. 2007).

Ultimately, although no one in the medical community can absolutely ascertain whether the oil was the reason Lorenzo plateaued and stayed alive as long as he did (he eventually succumbed to aspiration), Lorenzo made medical history by living until age 30 and was the oldest person living with the cerebral form of ALD. In the end, Lorenzo's Oil is seen as an effective prophylactic therapy when combined with genetic screening.

Genomics and ALD

It wasn't until 1993 that the gene mutation for ALD was identified in the ABCD1 gene—one year after the film, *Lorenzo's Oil*, was released (Engelen and Kemp 2013). Since then, no significant pharmacological treatment other than Lorenzo's Oil has been promoted, but the identification of the ALD gene in 1993 led to the ability to do genetic testing, and now newborn screening for ALD. From 1993 until the early 2000s, most of the ALD research involved further study of its genetic aspects, and Hugo Moser established an important ALD database of patients with confirmed ABCD1 mutations (a disease registry) that helped in understanding the genetics of ALD further (Moser et al. 2001). ALD was found to be among a group of distinct genetic diseases in which there is a problem with the peroxisomes, and thus, is said to be a "peroxisomal disorder". The most significant breakthrough in ALD is that it was added to newborn screening programs, which may permit ideal candidates for either Lorenzo's Oil, or more recently, gene therapy. I discuss the prospect of gene editing in Chap. 7, which may also have application here.

If teaching this film to pediatric trainees, it's important to review various clinical and ethical justifications for newborn screening. Generally, there is strong ethical justification for newborn screening for diseases with childhood onset (versus adult onset disorders). The criteria for newborn screening was developed in 1968, known as the Wilson-Jungner criteria (Wilson and Jungner 1968) and it's important that we screen for diseases that can be mitigated through preventative steps or actions. The ten criteria proposed by Wilson-Jungner have been endorsed by the World Health Organization (WHO) and the President's Council on Bioethics (President's Council on Bioethics 2008) for the applicability of any newborn screening program. ALD certainly meets all of these criteria:

1. The condition sought should be an important health problem.
2. There should be an accepted treatment for patients with recognized disease.
3. Facilities for diagnosis and treatment should be available.
4. There should be a recognizable latent or early symptomatic stage.
5. There should be a suitable test or examination.
6. The test should be acceptable to the population.
7. The natural history of the condition, including development from latent to declared disease, should be adequately understood.
8. There should be an agreed policy on whom to treat as patients.
9. The cost of case finding (including diagnosis and treatment of patients diagnosed) should be economically balanced in relation to possible expenditure on medical care as a whole.
10. Case-finding should be a continuing process and not a "once and for all" project.

The President's Council on Bioethics endorsed the original Wilson-Jungner criteria in 2008, highlighting that this classical criteria would mean the disorder for which mandated newborn screening is recommended "must pose a serious threat to the health of the child, its natural history must be well understood, and timely and effective treatment must be available, so that the intervention as a whole is likely to provide a substantial benefit to the affected child" (President's Council on Bioethics 2008). The American Council on Medical Genetics endorses mandatory newborn screening by stating: "Societies have an ethical obligation to protect their most vulnerable members, especially if these people cannot protect themselves. Newborns deserve the special protection afforded by mandatory screening for disorders where early diagnosis and treatment favorably affect outcome... The primary purpose of mandatory newborn screening is to benefit the newborn through early treatment" (President's Council on Bioethics 2008).

This film takes place in the 1980s, and demonstrates the psychosocial harm of diagnosing a genetic disease that does not have a treatment. We note the guilt the mother feels when she learns she is the carrier of this disease, but this case and film takes place prior to the announcement of the Human Genome Project, where identification of certain genetic diseases can be done. Now, in a post-human genome era, where whole genome sequencing and even gene editing is a reality, we may look upon Lorenzo's Oil in a decade from this writing as an antiquated therapy in light of true gene therapy available. That said, costs of genetic screening, gene sequencing, and in the future, gene editing, may mean that some can choose Lorenzo's Oil as a viable medical alternative.

Medical Research in Children

When Lorenzo Odone was born (1978), the National Commission for the Protection of Human Subjects of Biomedical and Behavioral Research had just published codified guidelines for research on children, (National Commission 1977). The Commission developed guidelines that were informed by the Nuremberg Code (1947), published in 1949 by the U.S. government (U.S. Government 1949), which stated that no research on children was ethical since all research required consent by a competent individual; past abuses in research by U.S. institutions, including the Willowbrook trial, which was mentioned in a famous "whistle-blowing" paper by Henry Beecher (1966), discussed more in Chap. 10; the success of the polio vaccine trials involving children (Carroll and Gutmann 2011); and finally, a famous debate regarding non-therapeutic research in children between two philosophers, Paul Ramsey and Richard McCormick (McCormick 1974; Ramsey 1976), known as the "Ramsey-McCormick Debates", which was solicited by the Commission itself (Carroll and Gutmann 2011). In this debate, Ramsey argues that the Nuremberg Code should be followed with respect to children being excluded from all non-therapeutic research that did not directly benefit them. "Ramsey proffers an unyielding thesis: Children, who cannot consent and thus cannot enter into covenants of loyalty, can never be used as research subjects. Only when the child is a prospective beneficiary of an investigative procedure may research be joined to that child's care" (Jonsen 2006).

McCormick argued the opposite: that children should be allowed to participate in research for the sake of the net social benefit that research provides, and that they are part of society and the research community. "McCormick countered with the thesis that parents' proxy consent for their children to participate in research could be ethically valid on the presumption that it was based, not on what the child would wish for himself or herself (the usual justification for proxy consent), but, more deeply, on what the child ought to wish" (Jonsen 2006).

According to one of the Commissioners (Albert Johnson), the Commission wrestled with the question: "Should persons incapable of consent ever be research subjects, even if the harm is minimal and there is prospective benefit?" (Jonsen 2006). Ultimately, the report, *Research Involving Children* (National Commission 1977) calls for a hybrid of protecting children, and also allowing them to participate in research—even nontherapeutic—recognizing that barring all pediatric research would mean that progress in finding cures and therapies for pediatric-related medical conditions would be impeded by overly rigid restrictions. Thus, in 1977, it was ultimately recommended that children aged seven years and older must assent to IRB-approved non-therapeutic research, which would presumably only approve research that offered "minimal risk" while their parents can give permission. As for therapeutic research, if it met the beneficence standard and the child was the direct "prospective beneficiary", then there was no ethical problem with enrolling the child, except if a child did not assent. This paved the way for the American Pediatric Association's guidelines on Pediatric Assent and Parental Permission (Committee on Bioethics 1995). Children are thus required to assent to the extent that they can, while refusals should be taken very seriously. In the Odone case, Lorenzo was clearly enrolled in a pediatric research protocol that met the criteria for *therapeutic research*, or even "innovative therapy" (see under Clinical Ethics Issues). It is not clear whether the oil was ultimately therapeutic or even beneficial for Lorenzo, but since there was no other standard therapy, and the oil had the *potential* to directly benefit Lorenzo, it did meet the standard for "ethical research" based on both *Research Involving Children* (National Commission 1977) and *The Belmont Report* (National Commission 1979), which outlined four principles for ethical research. I discuss the Belmont Report in more detail in Chap. 10.

When teaching this film, it's important to point out the distinction between therapeutic pediatric research and *non-therapeutic* research as well as ethically defensible goals of pediatric research (Diekema et al. 2006). Non-therapeutic research requires there is no more than "minimal risk" (see under Clinical Ethics Issues) while therapeutic research only requires the criteria of beneficence is met for the enrolled subject—that s/he stands to directly benefit from a research protocol that has the potential to maximize clinical goods and minimize clinical harms. So, did Lorenzo's Oil meet that standard? The protocol did because of the potential of the therapy to be of great benefit—which took a very long time to sort out. As for the outcome for Lorenzo, specifically, not so much. It all depends on whether we think we prolonged Lorenzo's suffering. Extending a life without quality may not be considered a benefit to many—and certainly was not to the Brazeal family, discussed above. There is thus a difference between judging the protocol and judging the decisions of the parents. Several open-label trials on Lorenzo's Oil, in fact, failed to show that it improved neurological or endocrine function, but there was never a proper placebo-controlled clinical trial done, either. Hugo Moser wrote in 2001 (Engelen and Kemp 2016): "It is our view that Lorenzo's oil therapy is not warranted in most patients who already have neurologic symptoms. The clinical benefit of Lorenzo's oil is limited at best".

In the context of no other treatment other than palliative and hospice care as the standard, it's hard to say that the Odones enrolled Lorenzo in "unethical" pediatric research, but there are also problems in this case, as the Odones were also in the role of clinical investigators, and thus the researcher being the parent of his subject (Lorenzo) created a conflict of interest of some magnitude. It's also worth addressing whether Lorenzo assented to being enrolled in the multiple trials that preceded, and proceeded from, Lorenzo's Oil. Any discussion about Lorenzo's assent should give rise to split opinions on the matter, but given that Lorenzo had begun to communicate with blinking and finger movements, we can probably state that Lorenzo assented *eventually* to his innovative protocols, but not able to assent when he was younger. It still remains unclear whether the outcomes would have been any different if Lorenzo had not taken the oil, as Moser suggests that he was among a group of anomalous cases. What was accomplished in this case, however, were broader research goals—that investigating a potential therapy for this form of ALD might benefit many future children, which is how we define "medical progress".

Clinical Ethics Theme

When teaching this film, the main ethical framework in pediatric ethics to emphasize is "the best interests of the child" concept, which really translates into an assessment of what would constitute a beneficent care plan in this case. Do the parents make decisions that are in the best interests of their child, or their own best interests? Regardless, the outcome in this case is that their decisions wind up being best for science, and hence, "somebody else's kid". Other concepts in this film have to do with "pathological grieving"—which is a term we use to describe an unhealthy grieving process, which may lead to decisions that are not in the child's best medical interests after all. However, even if the parents in this film were, in fact, contributing to the fund of knowledge, and volunteered their child to risk for a future child, is that a violation of beneficence? As discussed under the Medical Research section, there is consensus that children could be enrolled in even non-therapeutic research so long as the risk was "minimal", but who defines "minimal" and what does that mean? Some standards of minimal risk define it as no greater than the "risk of daily life"—but some lives are riskier and harder than others (Freedman et al. 1993).

When looking at this case from purely a *clinical ethics* perspective that addresses whether decision-making was specifically in *Lorenzo*'s best medical and psychosocial interests as a patient, I argue the answer is No. The standard of care that was most beneficent, in which suffering was reduced and not prolonged in the face of a devastating, degenerative condition, was probably what the Brazeale's chose: a palliative/hospice care approach that made the child comfortable. Even in the absence of Lorenzo's Oil, the Odones insisted on aggressive care, and in the film, there were many "cringe worthy" moments for any pediatric healthcare expert. Three scenes stand out in this regard: (1) "fly to baby Jesus"—where Lorenzo, in a PICU, is choking endlessly on his saliva, which is being continuously suctioned. Wondering how he could endure this for so many hours, his mother realizes that her child may be "afraid" to die for fear of disappointing her, and she tells him it's okay for him to let go and die, and "fly to baby Jesus". Although he rallies, it should have been a clear sign that goals of care have to change. Most pediatric ethicists in these cases would challenge the parents' insistence on continued aggressive care and recommend no further interference with a natural death (and may potentially begin family goals of care conferences). (2) "I'm not comfortable in this situation". In this scene, a homecare nurse with profound moral distress over continued aggressive interventions that stop a natural death, confronts the mother (Michaela Odone). She tells her: "Look, I'm not comfortable anymore with this situation; this child needs to be in a hospital with 24/7 supervision." The nurse is summarily fired, but is resigning anyway. Michaela thinks there's something wrong with the nurse, and not her own decision-making. In this situation, it would have been ethically permissible for the nurse to involve Child Protective Services. (3) "*What* mind?" A few months later, a more junior nurse, "Nancy," has no problem with her clinical duties, and even reading to Lorenzo (she does so mechanically), but takes umbrage when Michaela instructs her to show more emotion or feeing when she reads as "Lorenzo needs ministering to his mind" to which Nancy retorts under her breath: "*WHAT* mind?" And she has a point. At this stage, Lorenzo has been frozen motionless for months, and no one can tell whether any part of his personhood has survived demylenation. Reasonable people, at this point, might decide to allow natural death.

Finally, the Odones make an unusual "recruitment"—they bring to the United States a Comoros Island native they had befriended there, who Lorenzo was close to, and have him share in the caregiving of their son. How voluntary was this recruitment, and should this be hailed as creative, or unreasonable/irrational?

AND/DNR Orders

Even knowing the full outcome of this case, with Lorenzo living until age 30, and news that he was able to eventually reach a stage where he could understand and appreciate some parts of life, I would argue that the beneficent care plan in this particular case would have been a hospice/comfort plan with Allow Natural Death (AND) or Do Not Resuscitate (DNR). Such a plan would take care not to escalate a natural death, but to respect it, and not interfere with its occurrence. Eventually, Lorenzo in fact did die from aspiration. Lorenzo lived several years in a state of suffering with absolutely no quality of life with 24/7 bedside care and continuous saliva suctioning. Eventually, for reasons not entirely clear, he did "plateau" (around age 14) and seemed to regain some ability to swallow his saliva, blink and respond in some way. Was it worth it? It's difficult to defend this end point, given the years of suffering the child endured. Yet as a "research ethics" case, there is more to salvage morally.

"There has to be a life beyond Lorenzo": Pathological Grieving

Pathological grieving occurs when passage of time does not seem to alter the intensity of the grief, and/or the grief interferes with daily functioning of the bereaved. In this case, there is an unwillingness by the parents to accept the natural stages of a devastating diagnosis for their son. Anticipatory grief over the probability their son will die, combined with grief over the loss of their child as they once knew him, becomes both a motivating force for researching a cure, and also a "pathological" barrier to a more beneficent care plan for their son. In one powerful scene, Michaela's sister tells her that she has to begin to think about her own life, rather than sacrificing *everything* in service to her son: "There has to be a life beyond Lorenzo". In this case, the parents' lives cease altogether as every waking moment is about caregiving or researching a cure. This is romanticized in the film as parental heroism, but it's important to discuss with the class that it isn't normal; the Brazeal's approach, which some may criticize as "giving up" is actually a much more normal example of parental grieving in which they accept a diagnosis but are also weary of applying untested treatments that are still experimental. The Odone case is complex because the research goals and the care plan become entangled; the Odones have always been forthright that for them, researching a cure for their son is the care plan since the alternative is not acceptable.

Parental Decision-Making for Vulnerable Populations and the Harm Principle

The Odones were making decisions for Lorenzo as surrogates, and as such, were indeed ethically and legally permitted to enroll him in a research trial for an experimental therapy that stood to offer him benefit. However, it is not clear whether the Odones were making decisions *in the child's best interests*. At several decision points shown in the film, one might argue that it would have been reasonable to involve child protective services to investigate whether the decisions for Lorenzo were in his best interests, such as prolonging his suffering for an untenable goal of completely restoring him to normalcy. When parents refuse medical treatment for their children based on strong religious or other convictions, we would question parental decision-making and call child protective services to look into potential medical neglect. Here, we may have the opposite situation: parents who insist on prolonged aggressive therapy to extend life long enough to find a cure based on strong convictions, and pathological grieving. In either case, we could invoke the Harm Principle (Diekema 2004) as justification for seeking state intervention when parents either refuse life-saving medical treatment that carries a good likelihood of preventing significant harm, or parents insist on overly aggressive or heroic measures that also lead to harm. The Harm Principle, originally outlined by J.S. Mill, in his <u>On Liberty</u> treatise (1859) states:

> The only purpose for which power can be rightfully exercised over any member of a civilized community, against his will, is to prevent harm to others…The only part of the conduct of any one, for which [an individual] is amenable to society, is that which concerns others. In the part which merely concerns himself, his independence is, of right, absolute. Over himself, over his own body and mind, the individual is sovereign. (Diekema, 2004).

The Harm Principle establishes that a competent individual has complete autonomy over his/her own beliefs and actions so long as those beliefs or actions do not create a significant likelihood of serious harm to another person. If one's actions or decisions place another in harm's way, state intervention is justified.

Therapeutic Misconceptions

When teaching this film, it's important to address therapeutic misconceptions about the effect of Lorenzo's oil on Lorenzo himself by reviewing the medical literature and noting that the Odones did not cure their son, but ultimately served a greater good for science as it took a long time to assess the efficacy of Lorenzo's oil and ideal candidates. Lorenzo, as it turns out, was not the ideal candidate for this preventative therapy. The *New York Times* had reported at the time of the film's release (Kolata 1993):

> [R]esearchers and ethicists say they are deeply troubled by the way the movie so powerfully conveys the Odones' conviction that they invented a cure for Lorenzo's disease and that traditional science was no help when parents were faced with a suffering child... Researchers say the case of one child cannot prove anything about the oil, nor can the stories of other boys who have taken the oil and not yet become ill. The only proof of benefit, or lack of it, they say, comes from the scientific studies that the movie and the Odones so deplore.

> Scientists say that they, too, desperately want to cure deadly diseases. But they find they must fall back on the painstaking pace of science, the publication of results, and the disinclination to believe anecdotal evidence. Otherwise they have their own natural inclination to see benefit in new therapies, even when there is none. They say it can be hard to convince grieving parents of a child with an unrelenting illness that everyone loses when case histories are accepted as proof of benefit and the dispassionate evidence of science is discarded.

Conclusions

The Lorenzo Odone case and the treatment of "Lorenzo's Oil" depicted in this film is the perfect laboratory to examine concepts surrounding beneficence, pediatric ethics and the "best interests of the child". This film demonstrates the complexity of beneficence, as it is often unclear whether we are actually maximizing benefits, particularly when patients are also research subjects, which is covered in much more detail in the next chapter.

Film Stats and Trivia

- This movie is on TWM's list of the ten best movies to supplement classes in Science, High School Level.
- Don Suddaby plays himself in the film.
- Roger Ebert gave the film 4 out of 4 stars.
- The medical community found the film problematic because of therapeutic misconceptions of Lorenzo's Oil.

- Moser called Ustinov's portrayal of him an "abomination" (Maugh 2007).
- Lorenzo Odone died one day after his 30th birthday, May 30, 2008; Michaela died June 10, 2000; and Augusto died in 2013.
- Zach O'Malley Greenburg, who played Lorenzo, never acted again.
- In a 2014 interview at the Florida Film Festival, Susan Sarandon said that *Lorenzo's Oil* was originally conceived and shot with the intent that as Lorenzo got sicker and sicker, the movie would fade from color to black and white. However, the production ran out of the money needed to process the film in that way, and the movie ended up being in color from beginning to end.
- Laura Linney's film debut was in a small part.

From the Theatrical Trailer

Director: George Miller
Writer: George Miller, Nick Enright
Actors: Nick Nolte
Susan Sarandon
Peter Ustinov
Producer: Doug Mitchell, George Miller
Cinematography: John Seal
Editor: Richard Francis-Bruce, Marcus D'Arcy
Production Company: Kennedy Miller
Distributor: Universal Pictures
Release Date: January 1, 1993 (limited Release: December 30, 1992)
Runtime: 129 min.

References

Aubourg, P., C. Adamsbaum, M.C., Lavallard-Rousseau, F. Rocchiccioli, N. Cartier, N., I. Jambaque, C. Jakobezak, A. Lemaitre, F. Boureau, C. Wolf, and P.F. Bougneres. 1993. A two-year trial of oleic and erucic acids ("Lorenzo's oil") as treatment for adrenomyeloneuropathy. *New England Journal of Medicine* 329: 745–752. http://www.nejm.org/doi/full/10.1056/NEJM199309093291101.

Auburg, Patrick. 1994. More on Lorenzo's Oil (In Reply). *New England Journal of Medicine* 330: 1904–1905. http://www.nejm.org/doi/pdf/10.1056/NEJM199406303302615.

Augusto Odone. 2013. *The Economist*, November 16. http://www.economist.com/news/obituary/21589838-augusto-odone-world-bank-economist-who-derived-lorenzos-oil-treat-his-son-died-october.

Beecher, Henry K. 1966. Ethics and clinical research. *New England Journal of Medicine* 274 (24): 1354–1360.

Carroll, T.W., and M.P. Gutmann. 2011. The Limits of Autonomy: The Belmont Report and the history of childhood. *Journal of the History of Medicine and Allied Sciences* 66: 82–115. https://www.ncbi.nlm.nih.gov/pmc/articles/PMC2998285/.

Childs, Martin. 2013. Augusto Odone: Economist famed for "Lorenzo's Oil". *Independent*, October 30. http://www.independent.co.uk/news/obituaries/augusto-odone-economist-famed-for-lorenzos-oil-8913924.html.

Committee on Bioethics. 1995. Informed consent, parental permission, and assent in pediatric practice. *Pediatrics* 95: 314–317.

Diekema, Douglas S., and F. Bruder Stapleton. 2006. Current controversies in pediatric research ethics: Proceedings introduction. *The Journal of Pediatrics* 149: S1–S2.

DeCoursey, Thomas. 1985. Reagan proposal to cut medical research by 23%. *Los Angeles Times*, February 16, http://articles.latimes.com/1985-02-16/local/me-2930_1_research-works-biomedical-research-grant-proposals.

DeVera, Imelda. 2009. *Learning Guide to Lorenzo's Oil*. Teach With Movies. http://www.teachwithmovies.org/guides/lorenzos-oil.html. Accessed 26 Feb 2018.

Diekema, Douglas S. 2004. Parental refusals of medical treatment: The harm principle as threshold for state intervention. *Theoretical Medicine and Bioethics* 25: 243–264.

DiGregorio, V.Y., and D.J. Schroeder. 1995. Lorenzo's oil therapy of adrenoleukodystrophy. *Annals of Pharmacotherapy* 29: 312–313. https://doi.org/10.1177/106002809502900314.

Dutka, Elaine. 1992. The spark that gives 'oil' heat: Director George Miller follows his passion and gambles on long shot—A medical mystery. *Los Angeles Times*, December 30. http://articles.latimes.com/1992-12-30/entertainment/ca-2416_1_director-george-miller.

Engelen, M., and S. Kemp. 2016. History of ALD. ALD Database. http://adrenoleukodystrophy.info/clinical-diagnosis/history-of-ald. Accessed 19 Jul 2017.

Freedman, B., A. Fuks, and C. Weijer. 1993. In loco parentis: minimal risk as an ethical threshold for research upon children. *Hastings Center Report*. http://onlinelibrary.wiley.com/doi/10.2307/3562813/abstract.

Fumarola, di Silvia. 1990. Cosi' un padre diventa scienziato per salvare il figlio (Translation: "a father becomes scientist to save his son"). *La Republica*, February 27. http://ricerca.repubblica.it/repubblica/archivio/repubblica/1990/02/27/cosi-un-padre-diventa-scienziato-per-salvare.html?ref=search&refresh_ce.

Jonsen, A. 2006. Nontherapeutic research with children: The Ramsey versus McCormick debate. *The Journal of Pediatrics* 149: S12–4. https://www.ncbi.nlm.nih.gov/pubmed/16829235.

Kolata, Gina. 1993. Lorenzo's Oil: A movie outruns science. *New York Times*, February 9. http://www.nytimes.com/1993/02/09/science/lorenzo-s-oil-a-movie-outruns-science.html?pagewanted=all.

Lerner, Baron H. 2009. Complicated lessons: Lorenzo Odone and medical miracles. *Lancet* 373: 888–889.

Maugh, T.H. 2007. Hugo Moser, 82, neurologist's portrayal in 'Lorenzo's Oil' belied his real character. Los Angeles Times, January 26.

McCormick, Richard. 1974. Proxy consent in the experimental situation. *Perspectives in Biology and Medicine* 18: 2–20.

Moser, A.B., J. Borel, A. Odone, S. Naidu, D. Cornblath, D.B. Sanders, and H.W. Moser. 1987. A new dietary therapy for adrenoleukodystrophy: Biochemical and preliminary clinical results in 36 patients. *Annals of Neurology* 21: 240–249. https://www.ncbi.nlm.nih.gov/pubmed/2440378.

Moser, H.W., and J. Borel. 1995. Dietary management of X-linked adrenoleukodystrophy. *Annual Review of Nutrition* 15: 379–397. https://doi.org/10.1146/annurev.nu.15.070195.002115.

Moser, H.W., A.B. Moser, K. Hollandsworth, N.H. Brereton, and G.V. Raymond, 2007. "Lorenzo's oil" therapy for X-linked adrenoleukodystrophy: Rationale and current assessment of efficacy. *The Journal of Molecular Neuroscience* 33: 105–113. https://www.ncbi.nlm.nih.gov/pubmed/17901554.

Moser, Hugo W., Kirby D. Smith, Paul A. Watkins, James Powers, and Ann Moser. 2001. 131. X-linked adrenoleukodystrophy. In *Metabolic and Molecular Bases of Inherited Disease,* vol. 2, 8th ed, eds. Scriver, Charles R., Arthur L. Beaudet, William S. Sly, David Valle. New York: McGraw Hill.

National Commission for the Protection of Human Subjects of Biomedical and Behavioral Research. 1979. Belmont Report: Ethical Principles and Guidelines for the Protection of Human Subjects of Research (4110-08-M) (Report). Federal Register: U.S. Government.

National Commisson for the Protection of Human Subjects of Biomedical and Behavioral Research. 1977. Report and Recommendations: Research Involving Children (O77-0004). Federal Register: U.S. Government.

National Research Council and Institute of Medicine of The National Academies. 2005. *Report by the Committee on Ethical Issues in Housing-Related Health Hazard Research Involving Children, Youth, and Families.* Washington: The National Academies Press.

Odone, A., M. Odone. 1994. More on Lorenzo's oil. Letter to the Editor, *New England Journal of Medicine* 330: 1904. http://www.nejm.org/doi/pdf/10.1056/NEJM199406303302615.

Poulos, Alfred, and Evelyn F. Robertson. 1994. Lorenzo's oil: A reassessment. *Medical Journal of Australia* 160: 315–317.

President's Council on Bioethics. 2008. *The Changing Moral Focus of Newborn Screening.* Washington, D.C. https://bioethicsarchive.georgetown.edu/pcbe/reports/newborn_screening/.

Ramsey, Paul. 1976. The enforcement of morals: Nontherapeutic research on children. Hastings Center Report 21–30.

Rizzo, W.B., R.T. Leshner, A. Odone, A.L. Dammann, D.A. Craft, M.E. Jensen, and J.A. Sgro. 1989. Dietary erucic acid therapy for X-linked adrenoleukodystrophy. *Neurology* 39: 1415–1422. https://www.ncbi.nlm.nih.gov/pubmed/2682348.

Rizzo, W. B., P.A. Watkins, M.W. Phillips, D, Cranin, B. Campbell, B., and J. Avigan.1986. Adrenoleukodystrophy: Oleic acid lowers fibroblast saturated C22-26 fatty acids. *Neurology* 36: 357–361. https://www.ncbi.nlm.nih.gov/pubmed/3951702.

Rizzo, W.B. 1994. More on Lorenzo's oil. "In Reply". *New England Journal of Medicine* 330: 1905. http://www.nejm.org/doi/pdf/10.1056/NEJM199406303302615.

Ryan, Michael. 1991. Trying to will a miracle. *People Magazine*, October 24. http://people.com/archive/trying-to-will-a-miracle/.

Saxon, Wolfgang. 2000. Michaela Odone, 61, the 'Lorenzo's Oil' mother. *New York Times*, June 13. http://www.nytimes.com/2000/06/13/us/michaela-odone-61-the-lorenzo-s-oil-mother.html.

Storch, Charles. 1993. Lorenzo's oil stirs up troubled water for families of ALD. *Chicago Tribune*, January 22. http://articles.chicagotribune.com/1993-01-22/features/9303171049_1_augusto-odone-ald-united-leukodystrophy-foundation.

The Stop ALD Foundation. 2004. What is ALD? http://www.stopald.org/what-is-ald/.Accessed 26 Feb 2018.

Trials of War Criminals before the Nuremberg Military Tribunals under Control Council Law No. 10 (Vol 2) (Report). 1949 Washington, D.C.: U.S. Government Printing Office.

U.S. National Library of Medicine. 2018. Genetics Hom reference: X-linked adrenoleukodystrophy. https://ghr.nlm.nih.gov/condition/x-linked-adrenoleukodystrophy. Accessed 26 Feb 2018.

Vitello, Paul. 2013. Augusto Odone, Father behind 'Lorenzo's Oil,' dies at 80. *New York Times*, October 29. http://www.nytimes.com/2013/10/29/world/europe/augusto-odone-father-behind-real-life-lorenzos-oil-dies-at-80.html.

Wilson, James M., and Y. Gunner Jungner. 1968. Principles and practice of mass screening for disease. *Bol Oficina Sanit Panam*, 65: 281–393 (Also listed as: Wilson JMG, Jungner G. 1968. Principles and practice of screening for disease. Geneva: WHO. Available from: http://www.who.int/bulletin/volumes/86/4/07-050112BP.pdf).

Wuntch, Philip. 1993. Sarandon's strong will suits her in 'Oil'. *Dallas Morning News*, January 26. http://www.deseretnews.com/article/272093/SARANDONS-STRONG-WILL-SUITS-HER-IN-OIL.html.

Weighing Clinical Goods Over Clinical Harms: *Awakenings* (1990)

10

Awakenings is about a 1969 clinical trial involving the drug L-DOPA (levodopa), based on the nonfiction clinical memoir of the same name (1973) by Oliver Sacks, the principal investigator. Sacks documented his experiences of finding a "miracle drug" for a curious neurological condition, only to discover that the benefits of the drug were short-lived, which led to more questions about beneficence. The drug trial involved a small group of catatonic patients who were statue-like from *encephalitis lethargica*—the "sleepy sickness" pandemic (1916–1928) that had occurred when Sacks' patients were in their teens and 20s. When they start on L-DOPA, they "awaken" from their catatonic states to discover they have aged, and that they have missed half their lives. Some of them cope better than others, but ultimately, the drug's efficacy or therapeutic benefit wears off, eventually causing unacceptable side-effects. Most of the patients need to stop the drug, which returns them to their previous states. The practitioners involved in the drug trial with this group of patients are distraught over the ethical implications of the paradox of a miracle drug that eventually fails. If you can fit only one "beneficence" film into a clinical ethics curriculum, *Awakenings* is the film to use, which is a vehicle to discuss the principle of beneficence in both clinical and research settings. This film illustrates classic beneficence dilemmas about whether the drug does more harm than good in the final analysis, and the difficulty in weighing therapeutic benefits against potential harms, which include psychosocial harms. *Awakenings* asks questions about "innovative therapy" and demonstrates when an innovative therapy should become formal "research". It also looks at when it is ethically obligatory to end a clinical trial.

© Springer International Publishing AG, part of Springer Nature 2018
M. S. Rosenthal, *Clinical Ethics on Film*,
https://doi.org/10.1007/978-3-319-90374-3_10

The Dramatic History of *Awakenings*

This film is based on the autobiographical account of the L-DOPA trial as written by Oliver Sacks (1933–2015) in his 1973 book, Awakenings (Sacks 1973; Sacks 1983). Dr. Sacks was a consulting neurologist for Beth Abraham Hospital in the Bronx (1966–9), a chronic care hospital when he discovered 80 patients who were survivors of the sleepy sickness epidemic, who had become catatonic or "statue like" for decades until they started on the drug. (He calls the hospital in his book, "Mount Carmel" to de-identify the patients.) Dr. Sacks states on his website: "Awakenings came from the most intense medical and human involvement I have ever known, as I encountered, lived with, these patients in a Bronx hospital, some of whom had been transfixed, motionless, in a sort of trance, for decades" (Oliversacks.com 2017).

The book has gone through several editions and publishers over the years, but when it was first released in 1973 by the British publisher, Duckworth, Sacks was approached by documentary filmmaker, Duncan Dallas and a documentary of the same name was made, and shown, on British television. (Sacks and Dallas 1974; Sacks 1990). The library catalogue description of the documentary reads: "This 1974 Yorkshire Television documentary (never released in the United States) features interviews and dramatic footage of Dr. Sacks' original post-encephalitic patients who were awakened by L-dopa in the summer of 1969." (See: http://www. worldcat.org/title/awakenings/oclc/489851957). The documentary featured the patients discussed in the book, as well as Sacks himself. In 1974, the U.S. edition of the book was published by Doubleday. The paperback edition was published in 1976 by Penguin (England) and Random House (U.S.). In 1982, a third edition is published by Dutton in the U.S., and Pan Books in England. This edition captures the attention of Harold Pinter, who writes a play based on the book. Harold Pinter's play, "A Kind of Alaska" (Pinter 1982) was based on one of the patients in the book, (Rose R.) who was "stuck" in the year 1926 when she awoke, and could not grasp that it was 1969, even though there were many similar cultural shifts going on (see further). The play debuted in October, 1982 in London, at the Cottesloe theater, and the original cast included Judi Dench, as "Deborah" who was based on the patient, Rose R. from Sacks' book (Sacks 1990).

Yet another edition in hardcover was published in 1987 by Summit Books in the U. S., which ignites further passion for the content; 1987 was a pivotal year for dramatic treatments of the book, Awakenings. In 1987, John Reeves, a producer with the Canadian Broadcasting Corporation (CBC), approached Sacks to participate in a radio drama of the book, which aired in March. Then, later that year, a Chicago theater company (City Lit Theater Company) staged a separate production of "Awakenings" that was distinct from either Pinter's or the CBC's version (Sacks 1990).

The film, *Awakenings*, had been in discussion since 1979 with producers Lawrence Lasker and Walter Parkes. Lasker had read the book as an undergraduate student at Yale university, and wanted to make it a film, so he had optioned the book few years later. Lasker and Parkes had visited Sacks and the patients at Beth Abraham Hospital (called "Mount Carmel" in the book). The film was finally "green lighted" in 1987 and Steve Zaillian wrote the screenplay. The film was cast with Robin Williams in the role of a fictionalized Sacks, whose name in the film is "Dr. Malcolm Sayer", and Robert De Niro in the role of "Leonard", one of the first patients to try the drug, who is discussed in the book, and who Sacks felt "taught him the most" about the effects of L-DOPA in post-encephalitic patients. Sacks participated as a consultant on the script, and makes clear that while there are fictional and clinical departures in the film that are necessary for plot and character development, he was comfortable that the content was a good representation of what occurred, and that especially, Robert DeNiro's portrayal of Leonard and his symptoms ("Leonard L." in the book, whose real name was "Ed") was uncannily accurate—so much so that at one point he couldn't tell whether DeNiro was acting or having a neurological crisis. Of all the case histories featured in the book, the film zeroes in on the patient, "Leonard L.", but does touch on a number of other patients from the book. DeNiro spent many hours with the actual patient he was portraying (Sacks 1990).

Williams' portrayal of Sayer is all the more ironic, considering that Williams himself was put on L-DOPA when he was diagnosed with Parkinson's disease in 2014 before he took his own life (Smith 2015). The film was nominated for Best Picture, and was directed by Penny Marshall.

The research and development phase of the film followed a very similar pattern as *One Flew Over the Cuckoo's Nest* (see Chap. 6) in that the actors and crew went to the institution itself and spent many hours with the patients they were playing or filming. One of Sacks' patients ("Lillian T." in the book) is in the film as herself. Says Sacks (1990):

> The three of them – Bob, Robin and Penny – went to [Beth Abraham Hospital] several times to get the atmosphere and mood of the place, and most especially to see patients and staff who remembered the 'awakenings' of twenty years before. One evening especially moving ...was a gathering together of all of us – doctors, nurses, therapists, social workers – who had been at [Beth Abraham Hospital] in 1969, all of us who had seen, and participated, in the 'awakenings'...We realized again how overwhelming, how historic that summer had been....

> I showed the actors how Parkinsonian patients stood, or tried to stand; how they walked, often bent over, sometimes accelerating and festinating; how they might come to a halt, freeze, and be unable to go on. I showed them different sorts of Parkinsonian voices and noises...Parkinsonian *everything*. I counseled them to imagine themselves locked in small spaces, or to imagine themselves stuck with glue.

The Many Social Locations of *Awakenings:* 1920–1990

When teaching this film, there are multiple relevant time periods to review, but here I will focus on three: the timeframe of the patient's lives and the pandemic itself, which is when it struck the patients in their social lives (1916–1920s); the timeframe of the L-DOPA trial and its documentation by Sacks in his book, Awakenings (1969–1973); and finally, the timeframe of the film, *Awakenings* (1989–1990).

1916–1927: The Sleepy Sickness Pandemic and Its Victims

The patients portrayed in *Awakenings* are real, although they were all born more than a century ago. They are the age of the Baby Boomer's grandparents, who were born around or during World War I—when American Civil War veterans were still alive. The sleeping sickness pandemic that struck these patients in their youth occurred around the same time as the 1918 flu pandemic (fading around 1920), which decimated the U.S. population—particularly in urban centers. The flu pandemic killed more people than the Plague in the 1300s in terms of numbers (Barry 2005). *Encephalitis lethargica* was a separate pandemic that was eclipsed by the flu pandemic, but started making headlines in the press in the early 1920s. Anyone living during this period experienced and remembered the Spanish Influenza pandemic, as just about every household was affected; the 1918 flu killed more people in 24 h than AIDS did in 24 years (Barry 2005), resulting in 100 million deaths worldwide. Social distancing and wearing masks was just about the only preventative measure. This was also a time frame during and post-World War I (1914–18), and during and post the Russian Revolution (1917), which had enormous impact globally, and created a "Red Scare". Patients struck by the "sleepy sickness" were especially caught off-guard in this time frame.

The 1920s were the beginning of major social reforms that were being shed from nineteenth century: women were fighting for emancipation, and finally got the vote when the 19th Amendment passed, and took effect in 1921, which coincided with dramatic shifts in their fashions and social roles. Hemlines raised to the knees; hair was cut short; makeup changed; and underwear was liberated from the Victorian corsets; women also began to take part in society and the workplace, and they started to smoke. The Charleston overtook "ballroom" dancing, giving rise to the "flappers". Geographically, there was a migration from rural to urban centers at this time, which included a cultural audience for mass entertainment, particularly the beginnings of Jazz, hence the term "Jazz Age" was dubbed for the decade. Alcohol, which was banned by the 18th Amendment in 1920 created a "drug culture" of sorts in which millions were imbibing illegally as "Speakeasies" (underground bars/clubs) opened up around the country—the 1920s version of substance experimentation, bootlegging and bathtub gin. The decade was nicknamed the "Roaring 20s" because of all the massive social changes and liberating of social mores. It was also a timeframe where post-war, the consumer culture economy began (everyone was buying on credit), which created a false bubble that would later burst (1929) (Digital History 2017; Scholastic 2017). But most of these patients missed all of that. They missed experiencing the Great Depression; they missed experiencing World War II, the Atomic Age, and the Kennedy Assassination. They awoke from their trance-like states the same Summer as the Apollo 11 moon landing and Woodstock (see further). Oddly, from a social standpoint, the patients who "awoke" in 1969 were experiencing similar social conditions that existed in the 1920s: a post-war environment in which there was a reactive youth culture; a women's liberation movement (now it had gone way beyond the vote, with hemlines to the thighs); hair was worn loose and liberated from the coifed styles of the 1950s; and there was a prosperity and upwardly mobile middle class that were years away from the recessions that took place in the 1970s and 1980s. Patients who were in their teens and 20s in the 1920s parented the "Greatest Generation" and grandparented the Baby Boomers. In 1969, it was the generation in senior roles, only most of these patients missed their adult lives. When teaching, students will be of similar ages *a century later* to the patients who had the misfortune to succumb to the sleepy sickness pandemic. Students should ponder what it would be like to be "awakened" in 2069, for example, from the present day. In the case histories presented in Sacks' book, Awakenings, patients' birth years ranged from 1900 to 1924; 1908 was the most common birth year in these patient histories, and the average age the patients were struck with sleepy sickness was 12.8 years, ranging from age two to 21 (Sacks 1973).

During the 1920s, there were also intense political and civil rights problems, which included resistance to immigration due to an influx in new immigrants in the late nineteenth century, while the Ku Klux Klan—remnants of the Confederate army and their descendants—rose to four million strong, as a White Supremacist terrorist organization in the South. The infamous "Greenwood Massacre" took place in 1921, which was similar to the "pogroms" in Russia on Jewish towns in the late nineteenth century. In this case, white supremacists invaded this middle class African American community nestled in a suburb of Tulsa, Oklahoma, and essentially burned it down, and killed hundreds, leaving the town decimated (Greenwood Cultural Center 2017). On the immigration front, the quota system began in 1924, which was essentially a discrimination-based system.

The 1920s saw the development of radio and the commercial broadcast network. The invention of radio changed everything (as television would 30 years later). The very first commercial broadcast aired Nov 2, 1920, which was the Presidential Election returns (NPR 2002). By 1922, there were over 500 radio stations; in 1926, NBC and CBS launched as stations. In 1921, radio sales reached 12. 2 million, and increased to over 800 million by 1929. In New York City, there was an area known as "radio row" that was dominated by radio stores and accompanying parts, but by 1966, the year Sacks would begin to see his sleepy sickness patients, radio row was leveled for a new structure to be built, known as the World Trade Center. Other mass entertainment included motion pictures, and the creation of the "movie star".

Car ownership and mass car production began to take off in the 1920s as well, as a result of mass production. By 1925, two million cars had been produced, and transportation began to shift from horses to cars, and new paved roads began to transform the urban landscape. Electric lights, and a new country where most Americans now lived in cities made the 1920s a remarkable change decade.

In 1927, as the sleepy sickness began to disappear, Charles Lindbergh made his famous flight to Paris. The 1920s saw three Republican administrations: Warren

Harding (who died in office in 1923); Calvin Coolidge (1923–29), whose pro-business, anti-government ideas contributed to the seeds of the 1929 crash; and finally, Herbert Hoover (1929–33), the last Republican President until Eisenhower in 1953, who was unable to deal with the Depression.

Similar to the backdrop of the late 1960s, the younger generation rebelled in the 1920s while their parents were horrified at the cultural changes. The Roaring 20s and the Swinging 60s had much in common from a cultural shift perspective.

The Summer of 1969

When the L-DOPA trial began, it was the same summer the Apollo 11 mission (July 16–22) achieved its goal of putting a man on the moon on July 20, 1969. During this mission, the presidential aspirations of the third Kennedy brother ended with the "Chappaquiddick" incident (July 18), when Ted Kennedy drove a car into a river and saved himself instead of his female passenger who drowned (Mary Jo Kopechne). It was also the summer of the "Tate/LaBianca" murders (August 8–9) by the Manson Family, and the pivotal concert event, Woodstock (August 15–17). Few summers in the twentieth century had so many major historic events.

But of all the events, it was the moon landing that had the most profound effect on the scientific community. Essentially, most Americans were in "recovery" from one of the most tumultuous years historically in the twentieth century: 1968, the year both civil rights leader, Martin Luther King, and presidential candidate, Robert F. Kennedy, were assassinated on April 4 and June 6, respectively. The King assassination led to race riots across the country; at the same time, there were widespread demonstrations against the Vietnam War across American university and college campuses, and a violent Democratic National Convention that year as police and anti-war protesters faced off. The Apollo 11 mission, which put the first man on the moon on July 20, 1969, helped to send a message that scientific progress would save the country from its sociological demons. Set against this backdrop of the Apollo space program was a new television series that excited science buffs—a series about space exploration in a future where technology, human values and diversity thrive—Gene Roddenberry's *Star Trek* (1966–9), which helped to shape the technologies of the twenty-first century as its fans grew up to invent personal computers and smartphones. Thus, 1969 was a summer where possibility was in the air, and the perfect season for a "miracle drug", particularly in a timeframe of "drug culture", and when recreational hallucinogens were being promoted as social events.

What is now widely debated about the Apollo 11 mission was whether the level of risk to the human subjects—the astronauts—was acceptable, given the many technical glitches and failures of previous missions. As is often reported, an iPhone has more power than the equipment that flew human beings to the moon (Puiu 2017; Saran 2009). (Indeed, iPhone inventors were all *Star Trek* fans, trying to emulate the "tricorder"). But the Apollo program was actually framed more as a military mission where the astronauts were in service to their country and assumed risks in a similar fashion to ground troops. Nevertheless, the success of the mission, and the optimism that followed, inspired a new era in science, particularly in technology and engineering. Additionally, as the Nixon Administration began on January 20th of that year, there was a new sense of order in the country (Nixon was the original "law and order" candidate). Nixon won the 1968 election with 301 electoral college votes in an election against Hubert Humphrey (Democrat) and George Wallace, who was running as an Independent, and received 46 electoral college votes; no other Independent before or since had won any state's electoral college votes.

Nixon's approval rating that year was 67% (Coleman 2017) and the turbulence that had framed the 1960s since the JFK assassination, seemed to be coming to an end. In fact, Nixon enjoyed high approval ratings for most of his presidency until the very end, when it plummeted to 28%. However, Nixon was in fact pro-science; he devoted enormous resources to the issue of environmental science and established the Environmental Protection Agency, the *Clean Air Act* (1970), the *Clean Water Act* (1972); and the *Endangered Species Act* (1973). He also created what is now the National Cancer Institute as part of the NIH when he declared a "War on Cancer" with the *National Cancer Act* in 1971. Nixon had also proposed a comprehensive healthcare plan that would have been far more effective than the *Affordable Care Act* but could not get Democratic votes, despite Republican support (Climate Central 2012; Stockman 2012). Nixon did make some controversial decisions about science: he ended the NASA's human exploration program, citing resource allocation reasons, and in his second term ended the Scientific Advisor role that began under Truman (Logsdon 2014).

1970–3: The Book

As Oliver Sacks begins to write his clinical memoir about his unique summer experiments of 1969, another Apollo mission, coined "a successful failure" was mesmerizing the country: Apollo 13 (April 11–17), which was made into a feature film in 1995. Apollo 13 was essentially a "cursed mission" in which just about everything that could go wrong, did go wrong, and the mission, as planned, needed to be aborted. The astronauts were brought back safely due to the creativity and innovations made by the scientists on the ground. When the mission concluded, the dignity of the embattled astronauts with the bad luck of failed equipment emerged as the enduring NASA story. Similarly, the early results of the L-DOPA trial mirrored the "miraculous" moon landing, and the optimistic results seen with the first patients led to the enrollment of many more patients. But that particular drug trial needed to be aborted when it was clear that the drug was ultimately not beneficial. However, the L-DOPA trial patients, not unlike astronauts caught in limbo, were not just reduced to their side-effects; the dilemma, as discussed further was whether their quality of life—for good or for bad—was worth the pathological effects of the drug. Thus, as Sacks relays in his book, the L-DOPA trial, too, was a successful failure in the lessons that emerged from the trial regarding innovative therapy (see under Clinical Ethics Issues). By the time the book was published, another major event needed to be aborted: the Nixon Presidency. By the time Awakenings is published, Nixon is embroiled in the Watergate scandal; by the time the British documentary airs (1974), Nixon resigns.

1989–90: The Film

Principal production of the film, *Awakenings*, began in 1989, which marked the 20th anniversary of the L-DOPA trials, Apollo 11 and Woodstock. Optimism was infectious in 1989 as the Cold War thawed with the fall of the Berlin Wall by November of that year, and the beginning of the post-Cold war period. As the old Eastern bloc begins to dissolve, 1989 is also a year of possibility: Reagan ends his two terms and George H. W. Bush becomes the 41st President of the United States. But when the film is released December 22, 1990, it's amidst the Gulf War (August 2 1990–February 28, 1991), which though relatively short-lived, was the first major military conflict for the U.S. since Vietnam. In this war, the U.S. emerges victorious with relatively few casualties. On the NASA/space front, the Hubble Space Telescope also launched, and interest in space was renewed; the second successful launch of Gene Roddenberry's optimistic future was captivating space and science buffs again with *Star Trek: The Next Generation* (1987–94)—a series more popular than the original series (1966–69), with the USS Enterprise now under the ethical leadership of Jean Luc Picard. Economically, 1990 was not a great year; the mood in the country was somewhat down as a recession had begun. As noted in Chap. 3, in the early 1990s, a trend of white male "illness" films emerged, and *Awakenings* is perhaps no exception.

Synopsis

This film is the dramatized case of the real L-DOPA trial done by Oliver Sacks in 1969 at a New York City hospital. The film changes his name and centers around a young neurologist (Dr. Malcolm Sayer) who encounters a group of patients in a long- term care hospital who spent decades in strange, "frozen states"—all survivors of *encephalitis lethargica* "sleepy sickness" pandemic (1916–1927). (Some developed Parkinson's symptoms later). They were like human statues—they didn't move and were stiff. He notices that some of them may be aware, and reads about some promising research that suggests the drug L-DOPA may be useful. He tries it out on one patient (Leonard) with some success, and asks the hospital to support and help him fund a clinical trial with the drug for the rest of the patients with these symptoms. The trial has its ups and downs, and we learn about the risks and benefits of the drug, and whether it is appropriate to keep using. We start realizing there may be unintended consequences of the drug, and the film captures the moral distress of the neurologist, as he weighs what to do.

History of Medicine Context

There are three history of medicine contexts relevant here: the history of the encephalitis lethargica, or "sleepy sickness" epidemic (1916–1928); the 1918 flu pandemic, which completely overlaps with sleeping sickness; and the history of the drug, L-DOPA.

The 1915–27 Encephalitis Lethargica Pandemic: In the Shadow of 1918

Most patients had clear dates in their medical charts regarding the timing of contracting *encephalitis lethargica*, ranging from 1918–1927. Some patients were struck as young children (age 2.5), and some were struck as adults (the oldest being 21), but the average age that Sacks' patients got this unusual virus was around 13 years old. The patients varied with respect to when their neurological symptoms began; some showed signs immediately, and some did not have debilitating symptoms until later in adulthood, when the symptoms were then traced back to this particular outbreak. Sacks states in his 1973 preface:

> These patients are among the few survivors of the great [sleepy] sickness epidemic (*encephalitis lethargica*) of fifty years ago, and their reactions are those brought about by a remarkable new 'awakening' drug (levohydroxy phenylalanine, or L-DOPA). The lives and responses of these patients, which have no real precedent in the entire history of medicine, are presented in the form of extended case histories or biographies (Sacks 1990: xvii).

Also known as "von Economo disease" from the physician, Constantin von Economo, who first described it in 1917 along with Jean-Rene Cruchet (von Economo 1917; Cruchet et al. 1917; Sacks 1990), a pathologist, *encephalitis lethargica* is referred to as both "sleepy sickness" or "sleeping sickness". I prefer to use "sleepy sickness" here to distinguish it from Human African trypanosomiasis (WHO 2017), (transmitted by the tsetse fly), which is called "sleeping sickness", and which is a different disease. Sleepy sickness is an atypical form of encephalitis (inflammation of the brain), which can be caused by a virus or bacteria (i.e. bacterial meningitis), or even a parasite, and in some cases, may be an autoimmune disease. Prior to treatment for sexually transmitted diseases, it was commonly a complication of syphllis (Sacks 1990). In this case, *encephalitis lethargica* was a viral pandemic that appeared to have spread between 1915 and 1927 (some sources state 1917–1928), when it finally disappeared (Sacks 1990; Rosen 2010; Vilensky et al. 2007; Vilensky and Gilman 2011), and really hasn't come back as an epidemic. The virus attacked the brain, creating a "statue-like" existence of being motionless and speechless, but in some patients, it took several years before they became "frozen". The virus infected about 5 million people worldwide, and had an "acute" phase that killed about a third of its victims. For many of those who survived the acute stage of the infection, a "post-encephalitic" set of symptoms would eventually become debilitating, which resulted in severe neurological impairment and the "statue-like" state of existence. There was a lot of variation in how long it took Sacks' patients to become completely debilitated to the point where they couldn't function, but many lived with disabling symptoms while trying to carry on with their lives. In the case of "Leonard", born in 1920, he became completely debilitated by age 15, and his history documented being struck probably in early childhood. The film begins, in fact, with Leonard's strange symptoms in his pre-teens.

While "frozen" neurologically, the patients had awareness but couldn't move, emote, or care about events, and were essentially watching from a distance as though they were out of their bodies. One of Sacks' patients (Rose R., who is also somewhat depicted in the film) stated that she had an awareness of World War II, and the Kennedy Assassination, but did not "experience" it; when she "awoke" in 1969, she felt, and talked, as though it were still 1926, when she was a young and vivacious adult woman and "flapper" (Sacks 1990). From a medical history standpoint, sleepy sickness was a most bizarre illness that is still not well understood.

In the acute phase of the virus, patients suffered from a high fever, lethargy and sore throat, which could have been mistaken for the 1918 influenza (see further), but there were no pulmonary symptoms, and instead, vision problems (diplopia) and strange eye movements ("oculogyric crises"); delayed reactions (physical and mental); and sleep disturbances (upside down sleep schedules). Papers on strange behaviors of patients abounded (Stryker 1925). In some cases, full-blown catatonia would set in. Then there were the "Parkinsonian" symptoms—which were neurological symptoms typically seen in Parkinson's disease: tremors, tics (including vocal tics, such as repeating phrases, talking quickly, etc.), stiffness (especially in the neck), and upper body weakness; catatonia often co-existed with the Parkinsonian effects. So while these patients may have shared symptoms with those diagnosed with Parkinson's disease in the absence of any flu virus, Sacks' sleepy sickness patients were diagnosed formally, when he saw them originally in 1966, with "postencephalitic Parkinsonism", which was a diagnosis purely based on clinical observation and medical histories taken of the patients, but not based on the results of any specific diagnostic or confirmatory tests other than evidence of influenza antigens (Sacks 1990; McCall et al. 2008; Barry 2005); this was considered a unique condition specifically linked to victims of the 1916–28 *encephalitis lethargica* outbreak. There is even a hypothesis that Hitler showed post-encephalitic symptoms (Lieberman 1996). By 2008, there was still no further supportive evidence that Sacks' patients' symptoms were *definitively* linked to the sleepy sickness virus, but this is a case where epidemiology and case histories have been accepted as reasonable clinical proof (McCall et al. 2008). The only treatment for Sacks' patients appeared to be the innovative therapy of L-DOPA, and thereafter, other drugs used in Parkinson's disease, but as Sacks discovered in his L-DOPA trial in 1969, the drug generally failed to meet the criteria for "Beneficence" because it did not "maximize benefits" and "minimize harms"; it frequently did the opposite (see under "History of L-DOPA" further on).

The history of the "sleepy sickness" has to be understood in the context of a much bigger and more serious pandemic: the 1918 Great Influenza pandemic (aka

Spanish Flu) which was a global pandemic that probably started in a small Kansas town in the U.S. in the early part of 1918. According to clinicians at the time (Barry 2005: 232):

> The flu was ushered in by two groups of symptoms: in the first place the constitutional reactions of an acute febrile disease – headache, general aching, chills, fever, malaise, prostration, anorexia, nausea or vomiting; and in the second place, symptoms referable to an intense congestion of the mucous membranes of the nose, pharynx, larynx, trachea and upper respiratory tract in general and of the conjunctivae...[causing] absolute exhaustion and chill, fever, extreme pain in back and limbs...cough was often constant. Upper air passages were clogged."

There were also Ebola-like symptoms: 15% suffered from bleeding from the nose (epistaxis), and women bled from their vaginas due to uterine mucosa, not menstruation. "Blood poured from noses, ears, eye sockets" and when they died, "bloody liquid was seeping from the nostrils or mouths" (Barry 2005). For some flu victims, symptoms were very sudden and violent: sudden, extreme joint pain, chills and a high fever—almost an instantaneous reaction with no gradual buildup. In many flu victims, they could recall the exact moment they became sick, and many died within hours. According to Barry (2005): "JAMA reported 'One robust person showed the first symptoms at 4 pm and died by 10:am'."

The first victims of the flu were soldiers, who spread it through crowded conditions of various military bases and hospitals during World War I. For example, on October 4, 1918, over 100 men at Camp Grant died in a single day; in less than a month, 2800 troops reported ill in a single day.

> The hospital staff could not keep pace. Endless rows of men coughing, lying in blood-stained linen, surrounded by flies – orders were issued that 'formalin should be added to each sputum cup to keep the flies away' – the grotesque smells of vomit, urine and feces made the relatives in some ways more desperate than the patients. (Barry 2005: 216)

Based on the 20 selected case histories presented by Sacks in his book, Awakenings (out of 80 surviving patients he saw) the oldest patient was born in 1900; the youngest, born in 1924, and "Leonard", portrayed by Robert DeNiro in the film, was born in 1920. Anyone with a living memory of 1918 in the United States—particularly if they lived in an urban center such as New York or Philadelphia, in particular—experienced conditions similar to the plagues of the middle ages. Overcrowding and slums helped to spread the flu in cities, which were the same social conditions in Europe. As Barry notes, although the plague in the middle ages killed a much larger *proportion* of the population, in sheer body count, the 1918 pandemic killed more people in a year than the Black Death killed in in a century. As another parallel, the 1918 flu killed more people in 24 weeks than AIDS killed in 24 years. In Louisville, Kentucky, 40% of those who died were aged 20–24. The Great Influenza pandemic of 1918 was different than other flu epidemics because it seemed to target adults in the prime of life, rather than the very old or young, which was more typical. Half the victims were men and women in their 20 and 30s, or roughly 10% of all young adults alive in that time frame. The death toll so overwhelmed cities, there were coffin shortages; piles of dead bodies;

shortages of doctors and nurses (in Philadelphia, when the flu exploded in the population, all 31 of its hospitals were filled with flu patients). Relatives were either burying their own dead or wrapped them in cloth until "body wagons" (in trucks or by horse-drawn carriage) could pick them up. The deaths increased in 10 days from 2 deaths per day to hundreds of deaths per day. Most of Sacks' patients were children during this time; people isolated themselves, and children would see one neighbor after the other die and wonder who would die next. Everyone was wearing masks. Everyone had someone in the family who was dying or sick, and many households had several sick and dying. Common public health signs from 1918–20 would read "Spit spreads death" while ordinances were passed that made not wearing a mask a felony. Public events were cancelled, and city streets became empty and "ghost town" like. The pandemic didn't fade away until about 1920, and thus, there was considerable overlap between *encephalitis lethargica*, which started before, and ended after, the 1918 flu pandemic (McCall et al. 2008). There were reports in the newspaper about sleepy sickness throughout most of the 1920s. "The Mystery Malady"; "Sleepy Sickness Spreading"; "Tragedies of sleepy sickness: warped minds and broken bodies" (Sacks 1990:185).

Additionally, there were several 1918 flu victims that seemed to suffer behavioral and neurological after-effects. Some epidemiologists even wondered whether some of Sacks' patients' "post-encephalitic symptoms" were actually victims of the 1918 flu instead (Barry 2005). None of this speculation has ever been adequately resolved.

Doctor Shortages and the Status of Physician Education

During the timeframe of both epidemics—sleepy sickness and the Great Influenza—medical training and competencies were dramatically changing, too, in the wake of The Flexner Report (Duffy 2011), which sought to correct and standardize training of physicians. Several medical schools at the time did not meet basic standards in training physicians. This is discussed more in Healthcare Ethics on Film in the Professionalism and Humanism section.

The History of L-DOPA

Although Parkinson's disease was identified as a neurological disease, it wasn't until 1960 that symptoms were understood as problems with dopamine transmitters. This understanding of Parkinson's disease led to trials of L-DOPA (levodopa), which is a precursor to neurotransmitters, dopamine, noradrenaline and adrenaline. The first physician to work with L-DOPA in Parkinson's patients was George Cotzias, who published about his success in 1967, which is what led to Sacks' interest in it as an innovative therapy to potentially treat sleepy sickness survivors, whose symptoms were similar. L-DOPA can be manufactured in its pure form, and is a psychoactive drug that can cross the blood brain barrier, but dopamine cannot.

So when patients are administered L-DOPA, it is converted into dopamine by an enzyme, and then it increases concentration of dopamine in the brain for patients having difficulty making it on their own. Cotzias won the "Lasker Prize" (a U.S. version of the Nobel Prize) in 1969 for his discovery of L-DOPA as a treatment for Parkinson's. But in Sacks' patient group, L-DOPA had a short-lived effect, and its considerable side-effects were mostly intolerable for his patients, which included radical changes in behavior, such as tics and paranoia. In most patients, it did not meet the "beneficence" test because it did not offer any "therapeutic benefit; only pathological effects" (Sacks 1990).

In 1967–68, the cost of L-DOPA was $5000/lb; by 1969, the cost had come down to the point where a clinical trial with sponsors/donors, was feasible (Sacks 1990). The first patient was put on L-DOPA in March, "Leonard L", who is played in the film by DeNiro. The L-DOPA trial emerges during the "awakening" of clinical research ethics guidelines in the U.S. as well, but that is discussed under Clinical Ethics Issues further on.

By 1990, when the film was released, not much more was understood about either sleepy sickness or L-DOPA that hadn't been published by Sacks, and then revealed in the film. Eventually, even more traditional Parkinson's patients had similar responses to L-DOPA, in that it stopped working for many of them, too. However, the film spawned more research, and in the early 2000s, there were some further breakthroughs. It is now recognized that there is an L-DOPA "honeymoon" for anyone on L-DOPA (including Parkinson's disease, multiple sclerosis and amyotrophic lateral sclerosis) because once the dopamine deficiency is corrected by L-DOPA, there is still "neurological damage caused by metabolites of dopamine, which include dopachrome and other chrome indoles that are both hallucinogenic and neurotoxic" (Foster and Hoffer 2004). Although through the years, various theories on how to correct this have been proposed, "optimizing" therapy by prolonging the honeymoon period where the drug offers clear benefit remains challenging (Abbruzzese 2008). There has also been recognition that long-term use of L-DOPA leads to "Levodopa-induced dyskinesia (LID)" which remains an obstacle for long term success of L-DOPA (Figge et al. 2016). Ultimately, Sacks' detailed case histories and experiences with L-DOPA is what leads to a much more balanced view of the drug. Sacks notes in his book that his initial observations were seen as "overly critical" of the drug, and he had difficulty even publishing such observations in the peer-reviewed literature.

Clinical Ethics Issues

When using *Awakenings* for clinical ethics teaching, this film, at its core, is about the challenges of meeting the criteria for "beneficence" in either an "innovative therapy" context (when L-DOPA is first tried on Leonard) or in a clinical trial (when the early results prompted sponsoring a bona fide trial for other patients). Teasing out what constitutes "beneficence" should be the focus of the discussion, using the *Awakenings* case studies, and the mixed results of L-DOPA as the "laboratory" for examining these questions. This requires delving into the research ethics history of clinical trials in the U.S. as Sacks begins his work with his sleepy sickness patients at the "dawn" of more specific criteria for ethical trial design in the United States. Ultimately, the clinical ethics issues in *Awakenings* revolve around the ethical distinctions between innovative therapy and "research"; as well as the challenges to meet criteria for beneficence in a therapeutic trial, whether it's one patient, or 80.

1966–69: The Beecher Paper Era

When Sacks first begins to evaluate his sleepy sickness patients (1966–9), a ground-breaking "whistle-blowing" paper is published in the *New England Journal of Medicine* by Dr. Henry Beecher, who is concerned that there is not enough ethical oversight of the multiple clinical trials that are taking place all over the U.S. Beecher "outs" numerous unethical clinical trials in the paper (Beecher 1966) that he states violated the 1947 Nuremberg Code. Two years prior to Beecher's paper, the World Medical Association had just published the Declaration of Helsinki (WMA 1964) which largely echoed the Nuremberg Code, but called for the establishment of Ethical/Institutional Review Boards (IRBs or ERBs). This document has been revised many times over the years, but when Sacks is working with L-DOPA, it is the "inaugural" 1964 version of the document that is valid. When Beecher originally conceived of his paper, he sought to expose 50 studies in violation of the Nuremberg Code, but he needed to reduce the number to 22 "for reasons of space" (Beecher 1966). Some of the infamous studies exposed in the paper included the Brooklyn Jewish Chronic Disease Hospital and live "HeLa" cell study (HeLa cells were injected into elderly patients without consent); and the Willowbrook 1963–66 hepatitis studies, discussed more in Chap. 9. Ultimately, this paper leads to the formation of the National Commission for the Protection of Human Subjects in 1975, which then codifies research ethics guidelines in a publication known as *The Belmont Report* (National Commission 1979). These guidelines codified three core ethical principles: "Respect for Persons" which defines informed consent and protection for vulnerable populations, such as Sacks' patients. "Beneficence" which defined ethical trial design, which required that ethical enrollment ought to ensure there is a greater burden of benefits than harms so that one is maximizing benefits and minimizing harms. Finally, "Justice" dealt with distributive justice in research so that the burdens and benefits of research" (rather than access to healthcare, which is primarily what we're concerned about outside the research context) be evenly distributed across populations to avoid Tuskegee-like experiments again (see: Healthcare Ethics on Film). It's important to stress that based on what was available and known at the time, Dr. Sacks conducted an *ethical* trial of L-DOPA, and the exact same trial may still have been approved today by an IRB, given the same knowledge base. Today, patients are still placed on L-DOPA, even with practitioners knowing the drug has limits to efficacy beyond a certain time-frame. With informed consent, patients can choose to take such risks, however.

Innovative Therapy or "Research"?

The nuances between what is an "innovative therapy" versus a formal research study were not well defined in 1969, and for that matter, not even taught much when the film premiered in 1990. It's important to have this discussion in class, however, by providing core definitions. In 1986, Robert J. Levine, one of the commissioners and authors of *The Belmont Report* (National Commission 1979) authored the core text, Ethics and Regulation of Clinical Research (Levine 1986), in which he defines several terms and distinctions. Clinical research, he writes, is: "Research involving human subjects that is designed to either enhance the professional capabilities of individual physicians, or to contribute to the fund of knowledge in those sciences that are traditionally considered basic in the medical school setting." Levine also defines "innovative therapy": When uncertainties are introduced "by way of novel procedures ...in the course of rendering treatment." Thus, innovative therapy is nonvalidated practice—a therapeutic innovation for one patient. However, when there is "significant" innovative therapy (a therapy used on many patients, or something beginning to become used routinely), it should rise to become formal research.

Equipoise in Trial Design and Stopping L-DOPA

Concepts such as equipoise in trial design were certainly not well developed in the late 1960s, when Sacks was working with L-DOPA. However, it's important to discuss that Sacks' initial trial design started in "equipoise"—uncertainty whether L-DOPA was better than the standard of care (or "nothing" in this case). However, when equipoise was disturbed, and it became clear that L-DOPA was not potentially better than the standard of care, it was ethical to stop the drug, which is what Sacks did with most of his patients. This may lead into a rather nuanced discussion of equipoise, such as delving into core papers about the topic of "genuine uncertainty" or uncertainty principles (Fried 1974; Peto et al. 1976) in clinical trials. But when the film was released in 1990, another groundbreaking paper about ethics and clinical research was published in 1987 in the *New England Journal of Medicine* by Benjamin Freedman (Freedman 1987), who essentially works out many of the problems with equipoise and trial design. In this paper, a distinction is made between "theoretical equipoise" and "clinical equipoise" that helps researchers understand when it is ethical, or even obligatory, to design a randomized controlled trial. What Sacks experienced in the L-DOPA trial was the result of "theoretical equipoise". He had genuine uncertainty over whether Treatment A (L-DOPA) is better than Treatment B (standard of care, or nothing) in his patient population. His uncertainty became "disturbed" when he noticed his patient's side-effects, and he formed his own opinions, which made it difficult to continue the drug. However, there are many trials where different investigators have different results (based on different biases often), resulting in "clinical equipoise", which means there is uncertainty in the clinical community over whether Treatment A is better than Treatment B based on honest professional disagreement, and the individual investigator biases or hunches do not matter; only a large enough trial with enough statistical power matters, with the goal of *"resolving dispute in clinical community"* (Freedman 1987) because "progress relies on consensus within medical and research communities". When the film, *Awakenings*, was released in 1990, discussions about this drug trial coincided with much deeper ethical analysis of ethical trial design, based on Freedman's reframing of "equipoise".

Psychological Harms and Beneficence

During the timeframe of the L-DOPA trials and Sacks' writing of his clinical memoir of the trial (Sacks 1973), a Yale University researcher, Stanley Milgram was also busy writing his clinical memoir of a rather "shocking" trial that had involved a "deception" design to test how human beings respond to authority. Milgram's social psychology experiments measured the willingness of study participants to obey an authority figure who instructed them to perform acts that conflicted with their personal conscience. Milgram first described his research in 1963 in an article published in the *Journal of Abnormal and Social Psychology* (Milgram 1963*)* and then in greater depth in his 1974 book, Obedience to Authority: An Experimental View (Milgram 1974). Milgram's unique trial design recruited subjects asked to administer an electric shock of increasing intensity to a "learner". The subjects were told that the experiment was exploring effects of punishment (for incorrect responses) on learning behavior by administering "electric shocks". The subject was not aware that the "learner" in the study was *actually an actor*, and that no one was actually being harmed, and there was no electric shock involved. The results were astonishing, but left the deceived subject psychologically troubled and harmed. In the end, 60% obeyed orders to "punish the learner to the very end of the 450-volt scale, and no subject stopped before reaching 300 V. Milgram's data was used to help explain Nazi war criminal behavior, or other situations where good people "obeyed orders". Although Sacks may not have been aware of Milgram's work at all, the National Commission was, and for those reasons, "psychosocial" harms associated with research were included as part of the "beneficence" equation in research, and in *The Belmont Report*. A few years later, in 1971, as Sacks was still laboring over his Awakenings manuscript, another major psychological study was underway, known as the Stanford Prison Study—again, to examine limits to human behaviors and authority. Dr. Philip Zimbardo, the Principal Investigator, described the research as: "a planned two-week investigation into the psychology of prison life [that] had to be ended prematurely after only six days because of what the situation was and what the situation was doing to the college students who participated" (Haney et al. 1973). This reinforced that human subjects can be harmed by psychological situations, even when there is no medical treatment involved.

An important discussion to have when teaching the film, *Awakenings*, is the extent of the psychological harms involved when adults "awaken" to discover that they have missed half of their lives. I often start my *Awakenings* class with the question: Had the drug worked without side-effects, are there any harms? This forces thoughtful reflection on whether there are some psychosocial harms involved. The issue, as the discussion unfolds, is that on balance, the medical benefits would still outweigh such harms, but in the case of limited efficacy, it's difficult to make that argument in some patients.

Limits to Autonomy

Finally, when examining beneficence in this film, it's important to discuss whether it's a "check" on autonomy when patients want a therapy that does more harm, than good. We see this after the L-DOPA honeymoon wears off in Leonard, and Dr. Sayer tells him he can take the drug away against Leonard's wishes. When practitioners try to meet the criteria of beneficence, they may find that what, indeed, "maximizes benefit and minimizes harm" may not correspond to what patients want to do, who may want to risk more, or even refuse clearly beneficial therapies. In this film, the negative behavioral side-effects of L-DOPA, including paranoia, begin to interfere with decision-making capacity (see Part 2). It's important to discuss the distinction between the Principle of Autonomy, which assumes decision-making capacity, and the Principle of Respect for Persons, which carries a dual obligation to respect autonomous patients' wishes, and to protect *non-autonomous* patients who may make decisions that are "bad" for their overall health. However, when looking at quality of life, autonomous patient preferences guide decisions, or their advance directives, if they lose autonomy. We need to ask whether Leonard's quality of life with the negative side-effects are still "better" for Leonard than a "statue-like" affect. There are no right answers to this, of course, but practitioners are ethically justified in stopping therapies or limiting therapies that offer no benefit, or in their medical judgment, do more harm than good, or may do only harm and no good.

Autonomy "dilemmas" around therapy fall into three categories, and there may be a need to limit autonomy in the following circumstances. First, when patients request or refuse procedures or care plans that violate healthcare providers' obligation to maximize benefits and minimize risks (Beneficence), as in the case of Leonard insisting on continuing on L-DOPA when Dr. Sayer is concerned it is no longer effective, and seems only to be causing bad side-effects. The second circumstance is when a patient or surrogate seems to be violating the Harm Principle (see Chap. 9), and the healthcare provider feels the need to intervene. Finally, when patients require therapies they can't afford, or are not available to them, autonomy is necessarily limited, as in cases of resource shortages, such as solid organs. Ultimately, beneficence frequently clashes with autonomy because of different ideas about what is beneficial or risky from the practitioner and patient perspective. For autonomous patients, correcting misperceptions about therapeutic benefits through informed consent and addressing barriers to decision-making capacity and the U-ARE criteria can help to bridge the gap in these perceptions. For non-autonomous patients, ensuring there is an authentic surrogate decision maker who is making decisions based on patients' values and best medical interests is the solution.

Conclusions

Of all the films in this section, *Awakenings* is the film that examines beneficence from a research ethics and a clinical trial standpoint. This film can be used to discuss the criteria of beneficence for innovative therapies, trial design as well as therapeutic goals. From a clinical ethics standpoint, this film is about the "doctor's dilemma" with finding a suitable treatment for his suffering patients. Although students may wish to explore the issues raised in this film from the patient's perspective, including the autonomy issues raised surrounding Leonard's sense of confinement, it's important to emphasize how this film is really about ensuring drug safety and efficacy and not drawing conclusions too early before you see the full effects of the therapy.

Film Stats and Trivia

- Robin Williams, who plays Dr. Sayer, is prescribed L-DOPA years later (2014), when he is diagnosed with Parkinson's Disease.
- Penny Marshall originally wanted to cast Bill Murray in the role of Leonard, but she was concerned it would look like a comedy.
- Steven Spielberg was originally going to direct the film, but passed.
- Steven Zaillian, who wrote the script, came to Spielberg's attention when he was considering the film; he later hired him to write the script for *Schindler's List* (1993).

From the Theatrical Poster

Director: Penny Marshall
Producers: Walter F. Parkes and Lawrence Lasker
Screenplay: Stevan Zaillian
Based on: Awakenings by Oliver Sacks
Starring: Robert De Niro, Robin Williams, John Heard, Julie Kavner, Penelope Ann Miller

Music: Randy Newman
Cinematography: Miroslav Ondricek
Editor: Battle Davis, Jerry Greenberg
Production Company: Lasker/Parkes Production
Distributor: Columbia Pictures
Release Date: December 22, 1990
Run time: 121 min

References

Abbruzzese, G. 2008. Optimising levodopa therapy. *Neurological Science* 29: S377–S3779. https://www.ncbi.nlm.nih.gov/pubmed/19381767.

Barry, John M. 2005. *The Great Influenza*. New York: Penguin Books.

Beecher, Henry K. 1966. Ethics and clinical research. *New England Journal of Medicine* 274: 1354–1360.

Coleman, David. 2017. Nixon's presidential approval ratings. history in pieces. http://historyinpieces.com/research/nixon-approval-ratings. Accessed 19 Jul 2017.

Cruchet, Jean R., J. Moutier, and A. Calmettes. 1917. Quarante cas d'encéphalomyélite subaiguë. *Bulletins et Mémoires de la Société Médicale des Hôpitaux de Paris* 1917 (41): 614–616.

Digital History. 2017. Overview of the 1920s. http://www.digitalhistory.uh.edu/era.cfm?eraid=13. Accessed 19 Jul 2017.

Duffy, T.P. 2011. The Flexner Report—100 Years Later. The Yale journal of biology and medicine 84:269–276. https://www.ncbi.nlm.nih.gov/pmc/articles/PMC3178858/.

Figge, D.A, K.L. Eskow Jaunarajs, and D.G. Standaert. 2016. Dynamic DNA methylation regulates levodopa-induced dyskinesia. *Journal of Neuroscience* 36:6514–24. https://www.ncbi.nlm.nih.gov/pubmed/27307239.

Foster, H.D., and A. Hoffer. 2004. The two faces of L-DOPA: benefits and adverse side effects in the treatment of Encephalitis lethargica, Parkinson's disease, multiple sclerosis and amyotrophic lateral sclerosis. *Medical Hypothesis* 62:177–81. https://www.ncbi.nlm.nih.gov/pubmed/14962622.

Freedman, Benjamin. 1987. Equipoise and the Ethics of Clinical Research. *New England Journal of Medicine* 317: 141–145.

Fried, Charles. 1974. *Medical Experimentation: Personal Integrity and Social Policy*. Amsterdam: North- Holland Publishing.

Greenwood Cultural Center. 2017. Tula race riot. http://www.greenwoodculturalcenter.com/tulsa-race-riot. Accessed 19 Jul 2017.

Haney, Craig, Curtis Banks, and Phillip Zimbardo. 1973. A study of prisoners and guards in a simulated prison. Naval Research Reviews September:1–17.

Levine, Robert J. 1986. *Ethics and Regulation of Clinical Research*, 2nd ed. New Haven: Yale University Press.

Lieberman, A. 1996. Adolf Hitler had post-encephalitic Parkinsonism. *Parkinsonism Relat Disord* 2: 95–103. https://www.ncbi.nlm.nih.gov/pubmed/18591024.

Logsdon, John M. 2014. When Nixon stopped human exploration. *Planetary Society Blog*, October 28. http://www.planetary.org/blogs/guest-blogs/2014/1027-when-nixon-stopped-human-exploration.html.

Lost in Watergate's Wake: Nixon's Green Legacy. 2012. *Climate Central Blog*, August 8. http://www.climatecentral.org/blogs/richard-nixon-the-environmentalist-resigned-38-years-ago-today-14776.

McCall, S., J.A. Vilensky, S. Gilman, and J.K. Taubenberger. 2008. The relationship between encephalitis lethargica and influenza: a critical analysis. *J. Neurovirol* 14:177–185. https://www.ncbi.nlm.nih.gov/pmc/articles/PMC2778472/.

Milgram, Stanley. 1963. Behavioral Study Of Obedience. *Journal of Abnormal and Social Psychology* 67: 371–378.

Milgram, Stanley. 1974. *Obedience to authority.* New York: Harper & Row.

National Commission for the Protection of Human Subjects of Biomedical and Behavioral Research. 1979. Belmont Report: Ethical Principles And Guidelines For The Protection Of Human Subjects Of Research. (4110–08-M). (Report). Federal Register: U.S. Government.

OliverSacks Website. 2017. www.oliversacks.com. Accessed 19 Jul 2017.

Peto, R., M.C. Pike, P. Armitage, N.E. Breslow, D.R. Cox, S.V. Howard, N. Mantel, K. McPherson, J. Peto, and P.G. Smith. 1976. Design and analysis of randomized clinical trials requiring prolonged observation of each patient. I. Introduction and design. *British Journal of Cancer* 34:585–612. http://www.nature.com/bjc/journal/v34/n6/abs/bjc1976220a.html?foxtrotcallback=true.

Pinter, Harold. 1982. A Kind of Alaska. http://www.haroldpinter.org/plays/plays_alaska.shtml. Accessed 27 Feb 2018.

Puiu, Tibi. 2017. Your smartphone is millions of times more powerful than all of NASA's combined computing in 1969. *ZME Science,* May 17. http://www.zmescience.com/research/technology/smartphone-power-compared-to-apollo-432/.

Rosen, Denis. 2010. Review [of] Asleep: the forgotten epidemic that remains one of medicine's greatest mysteries [by] Molly Caldwell Crosby. Journal of Clinical Sleep Medicine 6:299.

Sacks, Oliver. 1973. *Awakenings.* London: Duckworth & Co.

Sacks, Oliver. 1983. The origin of 'Awakenings'. British Medical Journal (Clin. Res. Ed.) 287: 1968–1969.

Sacks, Oliver. 1990. *Awakenings.* New York: Vintage Books.

Sacks, Oliver, and Dallas, Duncan. 1974. *Awakenings* (documentary). Yorkshire Television.

Saran, Cliff. 2009. Apollo 11: The computers that put man on the moon. *Computer Weekly,* July 1. http://www.computerweekly.com/feature/Apollo-11-The-computers-that-put-man-on-the-moon.

Shaprio, Ben, and Joe Richmond. Radio Row: The neighborhood before the World Trade Center. 2002. *National Public Radio,* June 3. http://www.npr.org/programs/lnfsound/stories/020603.radiorow.html.

Smith, Nigel M. 2015. Robin Williams' widow: 'It was not depression' that killed him. *The Guardian,* November 3. https://www.theguardian.com/film/2015/nov/03/robin-williams-disintegrating-before-suicide-widow-says.

Stockman, Farah. 2012. Recalling the Nixon-Kennedy health plan. *Boston Globe,* June 23. https://www.bostonglobe.com/opinion/2012/06/22/stockman/bvg57mguQxOVpZMmB1Mg2N/story.html.

Stryker, Sue B. 1925. Encephalitis lethargica: the behavior residuals. *Training School Bulletin* 22: 152–157.

The United States Turns Inward: the 1920s and 1930s. 2017. Scholastic. http://www.scholastic.com/browse/subarticle.jsp?id=1674. Accessed 27 Feb 2018.

Vilensky, J.A., P. Foley, and S. Gilman. 2007. Children and encephalitis lethargica: A historical review. *Pediatr. Neurol* 37:79–84. http://www.pedneur.com/article/S0887-8994(07)00194-4/abstract.

Vilensky, Joel A., and Sid Gilman. 2011. Introduction. In *Encephalitis Lethargica: During and After the Epidemic,* ed. Joel Vilensky, 3–7. Oxford, ENG: Oxford University Press.

von Economo, Constantin. 1917. Encepahlitis lethargica. *Wiener Klinische Wochenschrift* 30: 581–585.

World Health Organization. Factsheet: Trypanosomiasis, human African (sleeping sickness). January, 2017. http://www.who.int/mediacentre/factsheets/fs259/en/. Accessed 19 Jul 2017.

World Medical Association, 1964. Declaration of Helsinki: Ethical Principles for Research Involving Human Subjects. Adopted by the 18th WMA General Assembly, Helsinki, Findland. https://www.wma.net/policies-post/wma-declaration-of-helsinki-ethical-principles-for-medical-research-involving-human-subjects/.